DISEASE AND DIAGNOSIS

Philosophy and Medicine

VOLUME 63

The titles published in this series are listed at the end of this volume.

DISEASE AND DIAGNOSIS
VALUE-DEPENDENT REALISM

by

WILLIAM E. STEMPSEY, S.J.
College of the Holy Cross,
Worcester, Massachusetts, U.S.A.

KLUWER ACADEMIC PUBLISHERS
DORDRECHT / BOSTON / LONDON

A C.I.P Catalogue record for this book is available from the Library of Congress.

ISBN 0-7923-6029-X (hb)
ISBN 0-7923-6322-1 (pb)

Published by Kluwer Academic Publishers,
P.O. Box 17, 3300 AA Dordrecht, The Netherlands

Sold and distributed in North, Central and South America
by Kluwer Academic Publishers,
101 Philip Drive, Norwell, MA 02018, U.S.A.

In all other countries, sold and distributed
by Kluwer Academic Publishers, Distribution Center,
P.O. Box 322, 3300 AH Dordrecht, The Netherlands

Material on pp. 28-32 is adapted from Robert M. Veatch and William E. Stempsey,
S.J., "Incommensurability: Its Implications for the Patient/Physician Relation."
Journal of Medicine and Philosophy 20 (1995): 253-269. Copyright 1995, Kluwer
Academic Publishers. Used with permission.

Printed on acid-free paper

Printed and bound in Great Britain by MPG Books Ltd., Bodmin, Cornwall.

To Mom and Dad

. . . . Thou, my Friend! art one
More deeply read in thy own thoughts; to thee
Science appears but what in truth she is,
Not as our glory and our absolute boast,
But as a succedaneum, and a prop
To our infirmity. No officious slave
Art thou of that false secondary power
By which we multiply distinctions, then
Deem that our puny boundaries are things
That we perceive, and not that we have made.

William Wordsworth
The Prelude

TABLE OF CONTENTS

PREFACE

The germs of the ideas in this book became implanted in me during my experience as a resident in clinical pathology at Boston University Medical Center. At the time, I had inklings that the test results churned out by our laboratories were more than scientific facts. As a philosophically unsophisticated young physician, however, I had no language or framework to analyze what I saw as a deep philosophical problem, a problem largely unrecognized by most physicians. The test results provided by our laboratories were accurate and of great practical importance for patient care. However, most of the physicians who relied on our test results to diagnose and treat their patients either did not have the time or interest to consider the philosophical issues inherent in diagnosis, or, like me, had inadequate means to further analyze them.

It was more than ten years later that I began doctoral studies in philosophy, and I was fortunate to find a faculty that was supportive of my efforts to address the problem. This book began as my doctoral dissertation in the Department of Philosophy at Georgetown University. I would like to acknowledge the assistance of my mentor, Robert Veatch, Ph.D. Our conversations during my Georgetown years led me in new and often fascinating directions. I would also like to acknowledge the help of Kenneth Schaffner, M.D., Ph.D., University Professor at The George Washington University, who provided helpful comments along the way, and Kevin Wildes, S.J., Ph.D., of Georgetown, who patiently read drafts of these chapters, offering many good ideas as well as support and encouragement. I am indebted to Angela Smith, my student at Holy Cross, for help in preparing the manuscript, and to Jeffrey Bernhard, M.D., of the University of Massachusetts Medical Center, for suggesting the apt epigraph from Wordsworth's *Prelude*.

CHAPTER 1

INTRODUCTION

This is a book about the philosophy of diagnosis. Diagnosis lies at the heart of the practice of medicine. It is part of the physician's everyday experience. Yet the philosophical presuppositions that underlie this common but complex practice are seldom considered by physicians.

The practice of medicine has long been said to be both an art and a science.[1] A common interpretation of this bit of professional wisdom holds that the science is exact knowledge: cold, logical, and unassailable. Science is the realm of objective facts, while art is just an expression of the subjective values of the individual physician. The art of medicine resides in the "bedside manner" of the physician. In diagnosis, the art of medicine might lie only in the skill and elegance with which the diagnostician gathers facts. In this common view, both science and art are necessary for the successful practice of medicine, but they are, in principle, separable.

This view, however, cannot survive the scrutiny of recent work in the philosophy of science. Medicine is an art not only because it requires the physician to engage in tactful social interaction with patients and colleagues, but also because medical research and interpretation of medical science are themselves arts. Research and science are arts in the sense that they require judgment about values as well as facts. The central thesis of this book is that values play a crucial and foundational role in delineating what we take to be objective fact. All our facts are value-laden, and our diagnostic facts are value-laden in a peculiarly complex way. In the search for medical facts, there is an art of value discovery and an expression of values that goes beyond what some claim to be a value-free scientific method.[2]

The naïve separation of fact and value often fails to recognize that at least some values are not merely subjective personal preferences, but are as objective as the alleged "scientific" facts in the practice of medicine. Some values ought to be recognized as objective values by everyone, just as facts ought to be recognized as facts by everyone. Furthermore, some values are objectively more valuable than others.

One way in which the fact-value dichotomy has been manifest in medical practice is in the division of the activities of the physician. A patient comes to a physician with an illness, where "illness" refers to a subjective experience of being unwell. In the biomedical model, the illness is held to be caused by a disease. The disease, in turn, has another cause, an infectious agent or a hormonal deficiency, for example. This cause logically and temporally precedes the disease, and the disease logically and temporally precedes the illness (Fabrega, 1979, pp. 554-556).

The physician's activities in trying to help the patient to recover from the illness, or to live with it, generally fall into three categories: diagnosis, prognosis, and treatment. The naïve separation of fact and value described above tries to make diagnosis a purely scientific process, an attempt to characterize the subjective illness of the patient as an objective disease process, and to find the underlying cause of the disease. As such, diagnosis is taken to be value-free. Once an accurate diagnosis is made, a prognosis can be known with varying degrees of certainty depending on the diagnosis. Only when the focus is shifted toward plans for treatment of the disease and management of the illness are value judgments made. For example, many diseases have several alternative treatments that tend to produce a similar outcome. The decision of which one to pursue will depend on value judgments of both patient and physician. Even the naïve fact-value separatist will admit that values come into play at this stage of scientific medical practice.

This separation of fact and value in medicine must now be questioned in light of the challenge of the philosophy of science of the past thirty years or so. The logical empiricists' project in the early part of the twentieth century held that any proposition that cannot be empirically verified is without meaning. All values are nothing more than preferences, and since preferences cannot be empirically verified, value statements have no meaning. With the abandonment of logical empiricism, some philosophers of science have turned their attention to the ways in which values are embedded in science. In so doing, they have forced a reconsideration of the very meaning of objectivity in science. A wide range of opinion exists today, from a new critical realism, which takes our theorizing to be about an independently existing reality, to social constructivism, which takes our world to be essentially constituted by our theorizing.

My project in this book will be to reevaluate the nature of medical diagnosis in light of this debate in the philosophy of science. The problem I intend to address is how facts and values interrelate in medical diagnosis. Medicine is a peculiar practice in which the blurring of the fact-value distinction becomes particularly evident. The danger in uncritically accepting a clear-cut fact-value distinction in diagnosis is that it can cause value-laden judgments to be seen as pure and unquestionable scientific facts. This can lead physicians to an unjustified practice of what we might call "diagnostic dogmatism." The dogmatic diagnostician fails to look for values that may be important to patients and so may ignore important elements of a patient's perception of health and disease. Diagnostic dogmatism is a type of scientism, an unfounded trust in what are seen as the indisputable facts of pure science. I will show that this view is an inadequate treatment of the complexities of the practice of diagnosis.

Different sorts of values may be of greater or lesser importance in classifying different diseases, and in coming to recognize these diseases as manifested in particular patients. But if some of the key values of diagnosis are objective values, we need not worry about turning the practice of diagnosis into a merely subjective and unscientific enterprise. We can still make objective judgments about our disease classifying and diagnostic activities. The theory that values are objective has a long philosophical tradition. Today, this tradition is usually referred to as "value realism," and I stand with this tradition.

The question of what constitutes a fact, with respect to both empirical observation and value judgment, will be critical. The thesis I will defend is this: The diagnostic process involves matters of fact about disease and our means to know disease, but all such facts and means are value-laden. Facts are not simply social constructions; they reflect a reality that is independent of our theorizing about it. Nevertheless, we cannot escape seeing disease through some constructed theory. Thus, we do shape all our knowing of the reality of disease by our theorizing, and we cannot avoid embedding values in our theories. Both fact and value are necessary in diagnosis; neither alone is sufficient. The key to understanding the interaction of fact and value in diagnosis lies in what I will call "value-dependent realism," which tries to mediate between the scientism of some types of scientific realism and the relativism of pure social constructivism.

That values are inherent in psychiatric diagnosis is more readily apparent than that values play an essential role in the diagnosis of somatic disease. Much of psychiatry deals with what is acceptable behavior, and values are evident in making those kinds of judgments. Some, such as

Thomas Szasz, have gone so far as to deny that mental illness is a disease at all. They would hold that mental illness is only a social judgment about people who have difficulties of various sorts in living everyday life, or who fail to conform to current societal values and norms of behavior. Szasz sharply delineates physical and mental illness, and rejects the medical model for describing mental illness.

> The concept of illness, whether bodily or mental, implies deviation from some clearly defined norm. In the case of physical illness, the norm is the structural and functional integrity of the human body. Thus, although the desirability of physical health, as such, is an ethical value, what health is can be stated in anatomical and physiological terms. What is the norm, deviation from which is regarded as mental illness? This question cannot be easily answered. But whatever this norm may be, we can be certain of only one thing: namely, that it must be stated in terms of psychosocial, ethical, and legal concepts (Szasz, 1970, p. 15).

I agree that psychiatric disease is inherently value-laden, but I will not deal specifically with the psychiatric realm in this book. My efforts will be directed at the somewhat more difficult problem of showing the importance of values in the concept and diagnosis of somatic disease. Szasz's dichotomy between psychiatric and somatic disease fails to recognize that somatic health, as well as psychiatric health, cannot be understood without reference to values.

In part one, I concentrate largely on the metaphysical issues raised by current philosophy of science and the fact-value debate. In chapter 2, I consider the challenges of the social constructivists and the scientific realists as they have attempted to fill the void in philosophy of science left by the abandonment of logical empiricism. I show how many of the insights of the social constructivists are helpful in understanding the normative nature of disease, but argue that these insights need not entail the relativism that many realists find in social constructivism. Value-dependent realism can incorporate these insights without rejecting the objectivity of either fact or value. Chapter 3 examines the distinction between fact and value. I argue that there can be no strict separation between fact and value because any determination of what constitutes the facts about any state of affairs will necessarily incorporate values, which I call *foundational* values. Some peculiarities of medicine especially blur the fact-value distinction.

In part two, I examine the concept of disease. Diagnosis purports to describe disease, and so any adequate analysis of diagnosis must consider the nature of disease. In chapter 4, I present several opposing metaphysical conceptions of disease, and then argue that the choice of one of these necessarily involves value judgments. Thus, the very concept of disease is value-laden. What this means is that any assertion that a particular illness is a disease involves not only facts, but values as well. I call these *conceptual* values.

In chapter 5, I consider the impact of the issues already raised on the classification of diseases. I analyze some established nosologies for their weighing of fact and value and argue that it is impossible to construct a classification of diseases without reference to values. Furthermore, nosologies include more than what might strictly be called a "disease." There are limitless observations that can be made about a person who is ill. Which of these are grouped together and considered to be constitutive of an entity worthy of nosological categorization will involve what I call *nosological* values. I discuss several examples of different types of disease.

In part three, I turn to diagnosis itself. In chapter 6, I consider the elements of diagnosis: history-taking, physical diagnosis, and laboratory and other testing by technological methods. The reporting and interpretation of the history of an illness and the interpretation of physical findings all involve values. I study certain problems in laboratory diagnosis, such as the fitting of continuous variables into a discrete taxonomy, and the statistical interpretation of tests, and I show how values are involved. I argue that *diagnostic* values are morally important at this stage of diagnosis because the benefit of information gained from a diagnostic procedure must be weighed against the potential harm to a patient from that procedure.

In chapter 7, I examine the process of diagnosis, including Bayesian, branching-logic and hypothetico-deductive models. I consider the role of expertise and intuition in medical diagnosis. Finally, I examine the use of computers in diagnosis and argue that at present humans are in no danger of being replaced by computers, not because computers cannot handle value judgments, but because the logic of diagnosis may be too complex and insufficiently precise for a computer to handle.

Thus, any time a physician makes a diagnosis, that diagnosis is embedded with four levels of values: foundational, conceptual, nosological, and diagnostic. Many sorts of values are included in each of these four levels. Some play important roles in several levels. Some, such as epistemic values, figure more prominently at the lower levels, while others, such as moral values, are most important at the highest level. Each succeeding level, however, incorporates the values of the lower levels. These particular four levels are necessary to understand diagnosis. A diagnosis, say, pneumococcal pneumonia, cannot be made without a diagnostic process to establish the existence of a set of physical signs in a patient and an etiologic agent, which is considered to be the cause of those signs. This process, in turn, depends on an already established nosology, a classification of types of diseases of the lung. Nosologies, in turn, are based on concepts of disease. We need to consider why we call pneumonia a disease before we classify it as a particular type of disease. Ultimately diseases are built from basic empirical facts and human values. Foundational values go into the facts about the basic signs that are observed in calling a physical state "pneumonia."

My conclusion is that when seen in the light of value-dependent realism, diseases are indeed real, even though they are in an important sense socially constructed from fact and value. The myriad values embedded in the conceptual elements and procedures that constitute the practice of diagnosis are an essential part of the diagnostician's discovery of the facts about a disease. With regard to disease and diagnosis, there are no facts without values.

PART ONE

FACT AND VALUE

CHAPTER 2

SOCIAL CONSTRUCTIVISM VS. SCIENTIFIC REALISM

I. INTRODUCTION

In part one, I will consider some issues that are foundational for our discussion of fact and value in medical diagnosis. I will give an account of the nature of a scientific fact, a value, and of the relationship between the two. This chapter is propaedeutic to our consideration of fact and value in diagnosis. It will also serve as an introduction to the view I call "value-dependent realism."

Because modern medicine is so grounded in science, it will be enlightening to consider these issues within the context of contemporary philosophy of science. How one understands the foundational concepts of fact and value in medicine will depend on one's particular view of the philosophy of science.

With the abandonment of logical empiricism by philosophers of science in the mid-twentieth century, two major competing views have come to the forefront in current philosophy of science: social constructivism and scientific realism. Although there is no consensus on which particular theory of the philosophy of science, if any, is the correct one, there has emerged a new consensus on several points.[1] Value-dependent realism is my attempt to draw upon some of these points of consensus in order to mediate between the extremes of scientific realism and social constructivism. It is in the context of value-dependent realism that I will develop my philosophy of medical diagnosis.

This realism-constructivism debate brings us face-to-face with some deep philosophical problems of metaphysical significance. The realist maintains that the world offers resistance to our theorizing about it. What is it in the world that offers this resistance? For the realist, it is the nature of the world itself, which is more or less described by our scientific theories. For the constructivist, it is only a particular worldview or ideology that offers resistance to our further theorizing. The value-dependent realist admits the importance of an objective reality, but recognizes the importance of values in shaping our descriptions of reality. What makes value-dependent realism a form of realism, and not just another idealist or constructivist theory, is its insistence that at least some

9

values are objective—as objective as the reality of the world is for the scientific realist.

II. COGNITIVE SIGNIFICANCE OF CONCRETE AND ABSTRACT ENTITIES

Diagnosis deals with both concrete and abstract entities. A carcinoma in the pancreas is an empirically verifiable entity. The tumor is readily observable (at surgery, at least); it is as real as the furniture in our homes. Yet to call this tumor a carcinoma, and to call carcinoma a disease brings us into the realm of abstract entities. One cannot observe diseases the way one can observe a particular tumor. Values (and perhaps facts, as well) are abstract entities. Values themselves cannot be observed in the same way we observe tumors. This is not to say that such abstract entities as values and diseases cannot be real, however. Even the scientific realist can admit the reality of such abstract entities.

The abstract entities with which we will be primarily concerned in our discussion of diagnosis are disease and disease classifications. Do diseases, which are abstract entities, exist in some realistic sense? Or is there nothing more than individual people who exhibit certain objective and subjective characteristics, which we have come to call diseases? This is an extension of the medieval debate between the realists and the nominalists with respect to universals.

In this section I will examine some of the issues surrounding the metaphysical status of these sorts of entities. This will prepare the way for a discussion of scientific realism and social constructivism and the mediation of value-dependent realism.

Logical Empiricism

Early in the twentieth century, the Vienna Circle, under the influence of Moritz Schlick, developed the philosophy known as logical empiricism or logical positivism. This was an attempt to push Hume's empiricism as far as it would go. Theoretical entities were taken to be validly introduced into a scientific theory only if they could be defined in explicit phenomenal or observational terms. The Vienna Circle was opposed to the introduction of metaphysical entities into science and philosophy. This view was broadened into a general theory of cognitive significance:

no discourse is meaningful unless it is given in terms of phenomenal language or terms that can be rephrased in phenomenal language. A verification theory of meaning was adopted. The meaning of a term, according to this theory, lies in the method of its verification. If some statement in ordinary language could not be empirically verified, it was concluded that that statement lacked any cognitive significance.

With regard to scientific theories, what emerged from logical positivism became known as the "received view" (Putnam, 1962, p. 240). Indeed, the received view outlived the logical empiricism that spawned it. The received view divides the non-logical vocabulary of science into two parts: observation terms, such as "red" and "touches"; and theoretical terms, such as "electron" and "gene." The observation terms signify qualities of things, qualities that are in principle publicly observable. Theoretical terms, on the other hand, refer to unobservable qualities and things. Statements are observational statements if they contain only observation terms and theoretical statements if they contain only theoretical terms. A scientific theory gets empirical meaning from the specification of meaning for the observation terms alone. Theoretical terms are to be defined in terms of observation terms, or phenomena. The logical positivist goal was to axiomatize science into theoretical statements that are given empirical content by relating them to observational statements, which refer to publicly observable phenomena, by means of explicit definitions called "correspondence rules" (Suppe, 1977, p. 12).

Rudolf Carnap's analysis of the problem of abstract entities illustrates how logical empiricism can incorporate abstract entities. Carnap asks us to consider whether or not there are things such as properties, classes, numbers, and propositions. He distinguishes two types of questions about the existence of such entities. There are questions about the existence of the objects within the linguistic framework we use to speak about the entities; these are the "internal questions." Then there are questions about the reality or the existence of "the system of entities as a whole"; these are the "external questions." Carnap takes the answers to internal questions, e.g., Is there a white piece of paper on my desk?, to be purely empirical. The concept of reality in internal questions is scientific and not metaphysical. Neither the ordinary person nor the scientist asks the external question about the reality of the "thing world itself." The philosopher, however, does. The realist affirms its reality and the subjective idealist denies it (Carnap, 1956, pp. 206-207).

The problem, according to Carnap, lies in the formulation of the question. He believes that the philosophers who think that questions regarding the ontological status of a linguistic framework must be settled before new language forms can be introduced are mistaken. On Carnap's view, language forms need no theoretical justification because they do not imply an assertion of reality. A statement about the alleged reality of a system of entities is a "pseudo-statement without cognitive content" (Carnap, 1956, p. 214).

How, then, do we designate abstract entities? In semantic analysis, there is little controversy about using certain language expressions to designate (name, denote) extra-linguistic entities when those entities (designata) are physical things or events. But some philosophers object to taking abstract entities as designata. Carnap gives the example of numbers as abstract entities. He bids us to consider a language L with a framework of numerical variables and the general term "number." Then, the statement

(1) "Five is a number"

is analytic in L. If suitable rules are established for the semantical relation of designation, then the following statement is also analytic.

(2) " 'Five' designates five"

But from (1) and (2) follows

(3) " 'Five' designates a number"

and so (3) is likewise analytic in L. Hence, *if* one accepts the framework of L, then (1), (2) and (3) must be acknowledged as true statements. So, abstract entities must be admitted as possible designata (Carnap, 1956, p 217).

Nominalists who doubt the existence of abstract entities are not debating the internal question, for within the language framework the existence of abstract entities is analytic, however trivial. Their doubts are about the external question. But Carnap sees the external question as only a question about whether or not to employ a particular linguistic framework, and not about the reality of the entities named by the framework. This decision is made on the pragmatic basis of its instrumental efficiency. For Carnap (1956, p 221), the external question about the existence of abstract entities, which is the question that gets debated by nominalists and universalists, is a question without cognitive meaning.

Problems with the Verifiability Criterion

Carnap's solution to the problem of abstract entities makes the deep question about the existence of these entities into a pseudo-question. Karl Popper, who, while remaining in the empiricist tradition, rejects much of the positivist position, cogently observes:

> The positivist dislikes the idea that there should be meaningful problems outside the field of "positive" empirical science—problems to be dealt with by a genuine philosophical theory. . . . For nothing is easier than to unmask a problem as "meaningless" or "pseudo." All you have to do is to fix upon a conveniently narrow meaning for "meaning," and you will soon be bound to say of any inconvenient question that you are unable to detect any meaning in it (Popper, 1992, p. 51).

For the logical empiricist, all scientific facts must be capable of being expressed as statements about sense experience. Popper, however, rightly points out that no scientific statement can avoid going beyond what we know with certainty on the basis of our sense experience. Every scientific description relies on *universal* names, symbols or ideas, and universals cannot be reduced to classes of experiences. Even the statement, "Here is a glass of water," contains universals, which cannot be correlated with any specific sense experience. The words "glass" and "water" denote physical things which exhibit a certain "law-like behaviour" (Popper, 1992, pp. 94-95).

Carl Hempel has pointed out the logical problems that beset logical empiricism's verifiability criterion for cognitive significance. Most fundamentally, statements of general laws and universal regularities will lack cognitive significance for logical empiricism. Being a stork and being red-legged are both observable characteristics, and being a stork does not entail being red-legged. So, "All storks are red-legged" is neither analytic nor contradictory, but it is clearly not deducible from any finite set of observation sentences. Furthermore, its negation, "There exists at least one stork that is not red-legged" is deducible from two observation sentences, "*a* is a stork," and "*a* is not red-legged." That the negation of a cognitively nonsignificant sentence yields a cognitively significant sentence is a further violation of the verifiability criterion (Hempel, 1965, pp. 102-105).

The verifiability criterion has further difficulties, which have been noted by Karl Popper. Popper is a kindred spirit of Hume in both his empiricism and in his take on the problem of induction. A widely accepted view of science is that it is an inductive method. Inductive methods begin with particular statements, such as observation reports from experiments, and move to universal statements, such as hypotheses and theories. The problem is, to take a familiar example, no matter how many white swans we observe, we cannot logically justify the conclusion that *all* swans are white. Popper rejects the Kantian notion that the principle of induction is *a priori* valid. He also rejects the doctrine that inductive inferences, although not strictly logically valid, are "probable inferences," which can attain some degree of reliability. If a certain degree of probability is to be assigned to statements on the basis of inductive inference, then this assignment must be based on a higher level principle of induction, and that principle on a still higher level principle, and so on in an infinite regress (Popper, 1992, pp. 27-30).

Popper's solution is to reject the verifiability criterion in favor of a falsifiability criterion. A scientific statement has meaning only if it is possible for the statement to be refuted through experience. This, according to Popper (1992, pp. 40-42), is the solution to Hume's problem of induction. Thus, the aim of science is not to show that scientific laws or theories are true, but rather to subject the laws and theories to experiment in attempts to prove them false. If one begins with a hypothesis, and rules it out by empirical testing, then one can use deductive logic to show that the hypothesis is false. Theories are merely more or less corroborated depending upon the extent to which they have been subjected to empirical testing and not yet been shown to be false (Popper, 1992, pp. 265-269).

Popper's theory seems to explain what physicians do when they make diagnoses. Data are gathered and a list of possible diseases that fit the empirical data (the differential diagnosis) is drawn up. Further data are gathered, which would rule out the various possibilities until only one diagnosis remains.

However, Popper has not succeeded in purging the logic of science of all inductive elements. If deductions are to be made from hypotheses, hypotheses must still be selected by some inductive method. Popper's method is to favor hypotheses that are easily falsifiable. But, those hypotheses that are not falsified are not taken to be proven, but only corroborated. As Wesley Salmon (1968, p. 26) shows, valid deductive inferences, though truth-preserving, are not ampliative. That is, a deductive conclusion cannot have any content that is not explicitly or implicitly present in the premises from which it is deduced. So, it is impossible to deduce from observation sentences any content that goes beyond that already contained in the observation sentences. Without induction, we would be without general theories and predictions, and science would be barren.

As Hilary Putnam points out, a scientist's acceptance of a law is a recommendation to others that they rely on it, often in practical contexts. If one only knows that a certain "provisional conjecture" has not yet been refuted, then one does not really understand anything. If we are to apply our scientific findings in future settings and anticipate success, then we cannot do without induction. Putnam (1991, pp. 122-123) rightly says: "The distinction between *knowledge* and *conjecture* does real work in our lives; Popper can maintain his extreme skepticism only because of his extreme tendency to regard theory as an end for itself."

Carl Hempel (1965, p. 133) argues that the demand for full verifiability or falsifiability as the standard for cognitive significance must give way to a standard of confirmability, that is, partial verifiability. This demand is properly applicable to whole theoretical systems rather than individual hypotheses. This suggests an affinity with W. V. O. Quine.

Quine, in a celebrated 1951 paper (reprinted in Quine, 1980, pp. 20-46), rejects the logical empiricist dogma of the analytic-synthetic distinction, along with the other empiricist dogma of reductionism, which holds that all meaningful statements are equivalent to logical constructs containing only terms that refer to immediate experience. While Quine wants to retain empiricism, his conclusion is striking. The unit of empirical significance is not the observation statement, but the whole of science. In our "web of beliefs," any statement, including one about what we take to be the most fundamental law of logic, is in principle open to revision.

The totality of our so-called knowledge or beliefs, from the most casual matters of geography and history to the profoundest laws of atomic physics or even of pure mathematics and logic, is a man-made fabric which impinges on experience only along the edges. . . . No particular experiences are linked with any particular statements in the

interior of the field, except indirectly through considerations of
equilibrium affecting the field as a whole (Quine, 1980, pp. 42-43).

What Quine shows is that no one statement or theory is absolutely
falsifiable; other beliefs in the web can always be altered in such a way as
to preserve the truth of the statement in question. Whether one decides to
change other beliefs or reject the statement under consideration will be a
pragmatic decision, and, I would add, one that may depend on one's
particular values.

The diagnostic process is not only an attempt to rule out false
diagnoses; it is essentially an attempt to explain an illness. A theory of
medical diagnosis is not an end in itself. Any such theory ought to give
satisfactory explanations for why we carry out potentially dangerous
procedures on people and give them drugs that can themselves cause
serious harm. Diagnoses are not merely generalizations about clusters of
empirical signs; they are warrants for treatment and grounds for
prognosis. It is the explanatory power of a correct diagnosis that is the
basis for rational treatment.

III. SOCIAL CONSTRUCTIVISM

By the 1960s, a number of philosophers of science were arguing that
philosophy of science cannot adequately be understood apart from a
consideration of the history and sociology of science. I will use the
commonly employed term "social constructivism" to refer to this
position, although I recognize that not all of these philosophers would
apply the term to their own particular philosophy. Common to these
positions are the assertions that scientific practice is essentially social in
nature and that scientific facts are as much constructed as they are
discovered. The work of Thomas Kuhn (1970a) has been considered to be
paradigmatic of social constructivism, although Kuhn's work is complex
enough to require some qualification of this label. Characteristic of social
constructivism is the view that the practice of science aims not at
discovering theories that are true, but at constructing theories that are
fruitful in supplying answers to questions that we find interesting. I will
examine Kuhn's theory after first presenting a theory that anticipates
much in Kuhn's position, and does so in the context of medical science.

Fleck's Thought Styles

In 1935, Ludwik Fleck, a Polish physician, observed that because medical problems deal with human life and well-being, they have an "individual and social significance such as is not directly within the provenance of any physical or chemical problem" (Trenn, 1981, p. 239).[2] He thus separated himself from philosophers who would see medical science solely as applications of the theories of the physical sciences. In the same year, Fleck published a book with the provocative title *Genesis and Development of a Scientific Fact* (Fleck, 1979).

Central to Fleck's position is the idea that scientific facts are not objectively given and therefore discovered by empirical observation. Rather, facts are social constructions, profoundly influenced by the culture in which they arise. He rejects the attempts of the Vienna Circle to establish absolute standards and criteria for knowledge (Fleck, 1979, p. 177, n. 3). For Fleck, what we take to be a scientific fact is profoundly influenced by the particular scientific milieu in which we find ourselves. Facts are the products of a "thought collective" (Denkkollektiv), which is a "community of persons mutually exchanging ideas or maintaining intellectual interaction." A thought collective gives rise to a "thought style" (Denkstil), which is a "special 'carrier' for the historical development of any field of thought, as well as for the given stock of knowledge and level of culture" (Fleck, 1979, p. 39). A thought style predisposes an individual member of a thought collective to perceive, feel, and act in certain culturally determined ways. Fleck defines thought style as "[the readiness for] directed perception, with corresponding mental and objective assimilation of what has been so perceived." The thought style "constrains the individual by determining 'what can be thought in no other way'" (Fleck, 1979, p. 99). A thought style is so fundamental for members of a thought collective that they are largely unaware of its influence on their thinking.

Fleck warns us that we cannot understand a foreign thought style in the terms of our own thought style. He cites an example of eighteenth century writings that describe how airy and fiery spirits cause a body to be lighter after a meal, and how the lack of such spirits cause a dead body to be heavier than when it was alive.

We have here a self-contained, logical system built up on a kind of analysis of feelings—at least on an identity of feelings. Yet it is a system completely unlike our own. Just as we do, these people observed, pondered, found similarities, and associated them. They set up general principles, and yet they constructed a system of knowledge completely different from our own. The "heaviness" in this last example is a concept totally different from that of our physical "weight." . . . *Our physical reality did not exist for them. On the other hand, they were prepared to regard many another feature as real which no longer has meaning for us* (Fleck, 1979, p. 127) (italics mine).

This leaves us with a problem. Different thought styles may be internally consistent, but may contradict each other. We have no way to determine which thought style is the correct one because there is no preferred frame of reference from which to decide such matters. Truth is a matter that can be decided only from within a thought style, and then only according to criteria that are determined by that particular thought style.

Differences between thought styles, however, may be greater or lesser. The greater the difference, the more difficult will be the communication of ideas. When two thought collectives share significant traits, communication is possible, although ideas are always changed somewhat as they are transferred between collectives. Very different thought styles may be used to explore any particular problem. Indeed, very different thought styles are more frequently used together than are very similar thought styles. For instance, a physician is more likely to study a disease from the bacteriological standpoint together with a historical standpoint than from the bacteriological standpoint combined with a purely chemical one (Fleck, 1979, p. 109-111). This not only provides a broader perspective on the issue being studied, but it avoids potential conflicts between similar thought styles, which are more likely to see the issue in conflicting terms.

Fleck characterizes the thought collective of modern science as consisting of an inner, esoteric circle made up of specialized and generalized experts, and an outer, exoteric circle made up of educated amateurs. The science of the non-experts is "popular science" or "textbook science," which is an artificial simplification that omits detail and controversy. This simplification allows for the appearance of a secure, consistent body of knowledge that can shape public opinion. Expert science, on the other hand, is constituted by "journal science" and "vademecum (handbook) science." Journal science is the science of the specialized experts. It is provisional and personal in nature. The problems it studies are sharply circumscribed, and its uncertain conclusions are submitted to the thought collective for scrutiny. The results of journal science are converted into vademecum science by the interchange of ideas throughout the thought collective. Vademecum science is not simply a collection of the ideas of journal science, for those ideas often conflict. Instead, the ideas are built into some orderly arrangement according to a plan that serves as a guideline for future research. The resulting vademecum science is the science of the general experts. Thus, the creation of a scientific fact is social in nature. The accepted position is one that is more exoterically conditioned than the provisional positions of journal science. Thus, the thought collective is democratic in nature, with the mass dominating over the elite in the final determination of what position becomes accepted, what will count as a fact, and what will be perceived as reality by non-experts (Fleck, 1979, pp. 111-125 passim).

Laurence McCullough (1981, pp. 258-259) rightly observes that in a thought style that emphasizes the predictive character of medical science, numbers and names take on a powerful significance. In diagnosis, laboratory findings tend to take on more significance than the findings of the patient's history and physical examination, which are seen as "softer" data. The interpretation of these quantitative measurements requires a sophisticated expertise. What results is a thought style in which a sociology of expertise plays a dominant role. McCullough sees that judgments are called for in the diagnostic process, and such judgments involve more than facts alone. A thought style determines which observations are seen as significant data, how they are seen, and how they are interpreted. Value judgments will necessarily be involved. I will say much more about this in subsequent chapters.

Kuhn's Paradigms

Thomas Kuhn provoked a revolution in philosophy of science over thirty years ago and has been the focus of much debate ever since. Kuhn's original position was laid out in 1962 in *The Structure of Scientific Revolutions*, a book in which he acknowledges that Fleck has anticipated many of his ideas. Kuhn's responses to critiques are found in a 1969 postscript in the second edition of the book (Kuhn, 1970a) and in numerous other writings.

Kuhn takes the practice of science to consist in relatively long periods of "normal science," which is based on a foundation of past scientific achievement. Normal science is characterized by a "paradigm," an accepted way of asking and seeking answers to questions. A paradigm delineates the sorts of questions that may be asked by scientists, limits the nature of acceptable answers to these questions, and specifies the methods that may be used in the process (Kuhn, 1970a, pp. 10-11). Working within a paradigm binds one to both methodological and "quasi-metaphysical" commitments. For instance, the Cartesian scientific paradigm assumed that the universe is composed of microscopic corpuscles. The work of the scientists in this paradigm is to explain all natural phenomena in terms of the activity and interaction of these corpuscles (Kuhn, 1970a, p. 41).

Paradigms are highly resistant to change. However, eventually an anomaly, a puzzle that cannot be solved within the paradigm, will arise. This crisis provokes a period of "scientific revolution" in which a new paradigm is sought to accommodate the anomaly and generate new puzzles and solutions (Kuhn, 1970a, pp. 66-91 passim).

Scientific revolutions are changes in worldview. A paradigm shift is somewhat akin to a Gestalt shift. There is no recourse to anything but observation itself in deciding which paradigm is correct. There is no privileged point of view from which to compare paradigms; the only alternative to interpretation of data within one paradigm is interpretation of data in some other paradigm (Kuhn, 1970a, p. 111-114). What this means is that paradigms are "incommensurable", that is, terms of one paradigm cannot be completely translated into the terms of another (Kuhn, 1970a, p. 148-150). Scientific progress is progress within a particular paradigm. There is no sense in which paradigm shifts carry scientists closer to truth (Kuhn, 1970a, p. 170).

Critique of Social Constructivism

Criticisms of Kuhn's philosophy of science have focused on four areas: ambiguity of the concept of a paradigm, the dichotomy between normal and revolutionary science, problems with incommensurability, and the charge that Kuhn's idealism makes him a relativist (Veatch and Stempsey, 1995, pp. 256-259).[3]

First, the notion of paradigm has been criticized as being ambiguous. Margaret Masterman documents twenty-one senses of "paradigm" in Kuhn's book. These senses, however, fall into three groups. There are *metaphysical* paradigms, the "set of beliefs," the "standard," the "new way of seeing," and so forth; *sociological* paradigms, the "concrete scientific achievement," the "set of political institutions," the "accepted judicial decision," and the like; and *artifact* or *construct* paradigms, the "actual textbook," the "actual instrumentation," or the "Gestalt-figure" (Masterman, 1970, pp. 61-65).

In response to such criticism, Kuhn has distinguished two senses of "paradigm." To the first, broader sense, Kuhn gives the name "disciplinary matrix." The disciplinary matrix is an ordered set of elements of various sorts that is the common possession of the practitioners of a professional discipline. The disciplinary matrix includes such things as symbolic generalizations, models, and exemplars. The second and narrower sense of "paradigm" is the exemplar. Exemplars are concrete solutions to problems that are well accepted by the professional community (Kuhn, 1977, pp. 462-463).

A second area of criticism of Kuhn concerns his distinction between normal science and scientific revolutions. Stephen Toulmin (1970, pp. 39-47), for example, does not find the notion of revolution to be very valuable as an explanatory concept. Instead, he sees Kuhnian revolutions as mere "units of variation" within the usual process of scientific change.

This aspect of Kuhn's thought, however, emphasizes the sociological basis of his philosophy of science. Sometimes revolutions (e.g., Copernican astronomy) are revolutions for everyone. At other times, they are revolutions for much smaller communities. There are such things as schools in science, groups that approach a particular subject matter from different points of view. It is these sorts of groups that ought to be regarded as the units that produce scientific knowledge. Some episodes will be revolutionary for only one such group, some for several, and very few for all (Kuhn, 1970b, pp. 249-254).

The third area of criticism has centered around the notion of the incommensurability of paradigms.[4] The most serious result of incommensurability, it is charged, is the inability to rationally compare theories belonging to different paradigms. Kuhn does not, however, think that incommensurability precludes the rational comparison of scientific theories. The view that it does rests on three misunderstandings (Hoyningen-Huene, 1990, pp. 488-490). The first is that Kuhn endorses "radical meaning change," or "total" or "radical" incommensurability. But Kuhn claims that he has never asserted that in a scientific revolution *all* concepts in the opposing theories have changed their meaning. He claims only "local incommensurability," that only a small group of concepts undergoes a radical meaning change in a scientific revolution (Kuhn, 1983, pp. 670-671).

The second misunderstanding is that there are no continuities between incommensurable theories. But Kuhn states that there are many continuities because a new theory must conserve much of the problem-solving ability of its predecessors or it will not be accepted by the scientific community (Kuhn, 1970a, p. 169).

The third misunderstanding is that incommensurable theories cannot rationally be compared at all. But because Kuhn holds that incommensurability is merely local, he can maintain that at least some of the consequences of the competing theories can be compared. Further-more, even if terms in the old theory cannot be mechanically translated into the terms of the new theory, a new conceptual vocabulary can still be learned. One can understand both Newtonian mass and Einsteinian mass even if the one sort of mass cannot be translated into the other without loss of meaning. Finally, if one rejects incommensurability due to value variance, one can globally compare incommensurable theories with respect to the values of simplicity, accuracy, fruitfulness, and the like.

The fourth area of criticism focuses on the perception of Kuhn's position as idealistic and ultimately relativistic. It appears that the reality of the world is entirely a social construction, and a construction that differs in different paradigms. Kuhn opens himself to these charges when he says:

> In a sense that I am unable to explicate further, the proponents of competing paradigms practice their trades in different worlds. . . . Practicing in different worlds, the two groups of scientists see different things when they look from the same point in the same direction (Kuhn, 1970a, p. 150).

This criticism is vitiated somewhat, however, if one continues to read what directly follows.

> Again, that is not to say that they can see anything they please. Both are looking at the world, and what they look at has not changed. But in some areas they see different things, and they see them in different relations one to the other (Kuhn, 1970a, p. 150).

However, one ought not to take Kuhn to be a realist. He removes any doubt about that in the Postscript where he says:

There is, I think, no theory-independent way to reconstruct phrases like 'really there'; the notion of a match between the ontology of a theory and its "real" counterpart in nature now seems to me illusive in principle (Kuhn, 1970a, p. 206).

For Kuhn, the world we see is not the world-in-itself, but only one out of many possible phenomenal worlds. While the concept of the world-in-itself may be arrived at through a process of subtraction of all the contributions of different phenomenal worlds, reality for Kuhn is always a particular phenomenal world, a particular way of viewing the world-in-itself. But the world-in-itself does offer resistance to our attempts to shape it. So, Kuhn is not a pure idealist, but occupies some middle road between realism and a social form of idealism (Hoyningen-Huene, 1993, pp. 267-271).

While there are substantial differences between Fleck and Kuhn, most notably, the absence of any notion of scientific revolution in Fleck's thought, the parallels in their thought are striking and significant.[5] While Kuhn does not make an explicit distinction between what Fleck calls a thought collective and a thought style, it would seem that a scientific community in Kuhn's view would be an example of Fleck's thought collective. Fleck's thought collective, however, makes more explicit the esoteric and exoteric circles that are important in the construction of knowledge.

Fleck's thought style parallels at least some of the ideas contained in Kuhn's "paradigm," especially the notion of disciplinary matrix. Both are socially determined products of a more or less defined community and both condition the kinds of questions that can be asked within a discipline and the kinds of answers that will be acceptable. Thus, both Fleck and Kuhn resoundingly reject the positivist idea that facts can be definitively specified only through logical structures and observation statements. Scientific knowledge is unavoidably conditioned by the group that produces it.

Thought styles, thought collectives, and paradigms, then, have slightly different focuses. None of these terms, however, has made explicit a value component. In this book, I will try to use these terms precisely as their authors have explicated them. I will use the term "worldview" as a convenient shorthand for an individual's (or community's) system of beliefs and values. This does not correspond exactly to Kuhn's use of "world view," but context should make clear any references to Kuhn's usage of the term. "Worldview," in my usage, is a system of beliefs and values of an individual (or community), but a system that is socially conditioned. It thus has much in common with Fleck's "thought style."

Characteristic worldviews typically arise in groups. Examples of such groups are the class of physicians as a whole, particular groups of specialists, researchers, philosophers of medicine, other health care professionals such as nurses and medical social workers, patients as a group, and particular subgroups of patients, such as those with a particular disease, or those of a particular ethnic background. Any individual may be part of a number of such groups, and so have several worldviews simultaneously. A surgeon might also be a philosopher of medicine and at the same time be a cancer patient. The surgeon, by virtue of his or her medical training, will tend to experience cancer in a way that is different from the way other patients experience cancer. Philosophical training might add another different slant to the experience.

Kuhn and Fleck both discuss the difficulty of communication between thought collectives, or scientific communities. To adopt Kuhn's language, there is at least partial incommensurability between knowledge claims in competing worldviews. The more unlike the worldviews, the greater will be the degree of incommensurability, and the more difficult will be the communication. I will, in subsequent chapters, show how patients and physicians, with different worldviews, tend to see different things when faced with an illness. Thus, there will be partial incommensurability in the diagnostic process as physicians try to "translate" a patient's subjective experience of illness into an "objective" disease classification.

Because both patients and physicians may simultaneously endorse more than one worldview, describing the physician-patient interaction in terms of worldviews becomes quite complex. A patient and physician may share a worldview, as when they hold the same religious beliefs. This may have important implications for their particular physician-patient relationship. The physician may even be afflicted with the same medical condition as the patient and so share that worldview. However, the patient *qua* patient and the physician *qua* physician will always play different social roles in the diagnostic encounter, and this may contribute to an important difference in worldviews.

What social constructivism gives us is a world in which science loses the strict objectivity it had for the logical empiricists. History and sociology take over a large part of the role formerly played by epistemology. Knowledge is always knowledge for a thought collective and the operative thought style determines what counts as knowledge.

Knowledge claims may or may not be accepted by different thought collectives, depending on how similar the thought collectives are with respect to thought style.

The insights of social constructivism serve as a foundation for a central claim of value-dependent realism: our understanding of diseases and our methods for classifying them and diagnosing them are necessarily fashioned by particular theories that arise in particular worldviews. In subsequent chapters, I will show how values are essential parts of these worldviews.

Social constructivism, however, does not give a very satisfying explanation of the nature of the world that "offers resistance" to our theorizing about it. If the world is fundamentally constituted by our theorizing, how can we explain the apparent match between our fundamental scientific theories and the way the world presents itself to us?

IV. SCIENTIFIC REALISM

Scientific realism, the major competitor of social constructivism in contemporary philosophy of science, focuses on the above question. The scientific realist claims that if a scientific theory has been intensely tested and shown to be empirically adequate, it is very likely to be true, truth being a correspondence of theory to the way the world is independent of our thinking about it. Richard Boyd describes scientific realism as having four central theses:

(i) "Theoretical terms" in scientific theories (i.e., nonobservational terms) should be thought of as putatively referring expressions; scientific theories should be interpreted "realistically."

(ii) Scientific theories, interpreted realistically, are confirmable *and in fact often confirmed* as approximately true by ordinary scientific evidence interpreted in accordance with ordinary methodological standards.

(iii) The historical progress of mature science is largely a matter of successively more accurate approximations to the truth about both observable and unobservable phenomena. Later theories typically build upon the (observational and theoretical) knowledge embodied by previous theories.

(iv) The reality which scientific theories describe is largely independent of our thoughts or theoretical commitments (Boyd, 1991, p. 195).

The social constructivist clearly rejects (iv), and most likely would also reject (i), (ii) and (iii), although these latter three might be accepted by the social constructivist if they are interpreted as applying to one particular Kuhnian paradigm.[6]

Scientific realists claim that the strongest evidence for their position is the instrumental reliability of scientific methodology. Scientific realism holds that scientific theories are approximately true accounts of the way the world, independent of our theorizing, actually is. Thus, scientific realism is defended as a scientific claim, a claim about how scientific theories work in the world. Boyd claims that this defense shows what is wrong with the social constructivist assertion that scientific methodology is so theory-dependent that science is more an invention than a discovery of truth. The constructivist claims that science is instrumentally correct because, in a sense, we have constructed our world in such a way that our theories must be successful (or they are rejected in a scientific revolution). The realist counters that our scientific methods are reliable because our scientific laws and theories are true, or approximately true, about a world we inhabit but do not create (Boyd, 1991, pp. 207-211).

The Scientific Image

What, then, is the nature of this world that we inhabit, but do not create? Scientific realism does not, unfortunately, give us the straightforward answer we might like to rebut the social constructivist. Wilfrid Sellars's version of scientific realism distinguishes the "manifest image" and the "scientific image." The "manifest image" is not simply the "original image," that is, our naive, pre-scientific conception of things in the world. It is, rather, the original image refined in two ways: empirically and categorially. By empirical refinement, Sellars means that a "disciplined and critical" framework gives rise to the images that the "perennial philosophy of man-in-the world" refines and endorses. Categorial refinement is an adjustment in the conception of what sorts of basic objects constitute the framework. The "perennial" philosophies, which include not only the Platonic tradition but also the "common sense" and "ordinary usage" schools, take the manifest image to be real (Sellars, 1991, pp. 6-14).

The "scientific image," by contrast, postulates imperceptible objects and events in order to explain correlations among perceptibles. Thus, the scientific image is derived from the "fruits of postulational theory construction." There are as many scientific images as there are sciences. To constitute *the* scientific image, one must telescope *some* of the "partial" images into one image. For example, objects of biochemical discourse may be equated with objects of the discourse of theoretical physics. This is to replace two theoretical frameworks with one theoretical framework with two levels of complexity. However, it is not to equate the two sciences, for it may be the case that the laws pertaining to the complex molecules of biochemistry cannot be discovered through the techniques and procedures of theoretical physics, but only through those of biochemistry (Sellars, 1991, pp. 18-21).

The manifest image and the scientific image compete in claiming to provide complete accounts of what exists. Sellars defends a scientific realism in which manifest objects are merely "appearances" of a reality that is constituted by systems of imperceptible particles. He considers the "knowledge objection" of the common sense philosophers:

We know that there are chairs, pink ice cubes, etc. (physical objects). Chairs, pink ice cubes are coloured, are perceptible objects with perceptible qualities. Therefore, perceptible physical objects with perceptible qualities exist.

However, he counters that this argument "operates *within* the framework of the manifest images and cannot *support* it" (Sellars, 1991, pp. 27-28). Sellars holds that even though the manifest framework successfully guides us in our everyday actions, the scientific framework is a more adequate *replacement* for the manifest image.

This sort of scientific realism has, on the surface, much in common with the practice of diagnosis in medicine. In general, the task of diagnosis is to redescribe a set of subjective symptoms and observed signs (phenomena) as a disease. It is the disease that the physician takes to be fundamentally real. The signs and symptoms that a patient brings to the physician are just phenomenal manifestations of the underlying reality, the disease. The disease is described within the scientific framework, and taken to be ultimately what is real.

This is not to suggest that symptoms are not real to a patient. They are just as real as tables and pink ice cubes. The scientific realist, however, sees them only as indications of a more fundamental reality, the disease, which is to be explained in scientific rather than experiential terms. This view has fostered great progress in the treatment of suffering patients, but it can also be destructive of the doctor-patient relationship. When physicians think as scientific realists, they can tend to be dismissive of a patient's complaint of symptoms for which no "scientific" basis can be found. The physician can unconsciously communicate that the patient's suffering is not real because it is not scientifically caused. The suffering, however, remains quite real to the patient.

Husserl's Critique

This sort of scientific realism is a manifestation of what Edmund Husserl (1970) sees as a more general intellectual crisis in Western thought. The crisis is precipitated by the idea of a "rational infinite totality of being with a rational science systematically mastering it." Galileo's "mathematization of nature" led to a radically new way of looking at nature itself: as a "mathematical manifold" (Husserl, 1970, pp. 22-23). In other words, scientific thinking (not only in the natural sciences, but in philosophy as well) *idealizes* the world. Galilean science refuses to accept the reality of the world as given in sense-experience; it tries to dig behind this crust and get at a mathematical structure of the universe. It is this mathematical structure that constitutes reality in Galilean science. Hobbes, for example, took phenomena to be in the subject, "only as causal result of events taking place in true nature, which events exist only with mathematical properties" (Husserl, 1970, p. 54).

With the mathematization of reality, reality becomes potentially understandable in its entirety, even if this is an infinite task of approximate "subsumptions" of empirical data under their ideal concepts. Husserl, however, argues that this view has no firm philosophical foundation. In fact, the new "philosophy *ordine geometrico*" coming forth as a new "naturalistic psychology" results in the paradoxical philosophical skepticism of Berkeley and Hume (Husserl, 1970, pp. 65-66). What actually drives Hume's philosophy is the problem of how the "*naïve obviousness* of the certainty of the world"—both the *everyday* world and the theoretical constructions built upon it—is to be made comprehensible. After unmasking Hume's naturalistic presuppositions, one comes to realize the *naïveté* of speaking about "objectivity" without appreciating objectivity's origin in "subjectivity as experiencing, knowing, and actually concretely accomplishing" (Husserl, 1970, p. 96).

Science is a human accomplishment. It presupposes the intuitive surrounding life-world, the *"Lebenswelt,"* "pregiven" for all in common. The practice of science "continues to presuppose this surrounding world as it is given in its particularity to the scientist" (Husserl, 1970, p. 121). The *Lebenswelt* is not to be confused with the objective-scientific world, although they are related. The *Lebenswelt* contains more than just objects. It also contains the ideal constructions of the sciences. But the objective-science world is "grounded" in the self-evidence of the *Lebenswelt*. The *Lebenswelt* is pregiven to the scientist. Yet as scientists go about their work of theory building, their work is pulled into the *Lebenswelt*, paradoxically building up the *Lebenswelt* on which it is grounded. We tend to distinguish scientific truth from life-world truth. We have a bias toward accepting scientific truth over life-world truth as real, forgetting about the dependence of the former on the latter. Husserl's phenomenological method is an attempt to remove this bias. Although I will not pursue Husserl's phenomenology any further here, I do take his trenchant critique to reveal a fundamental difficulty with scientific realism. The scientific realist presupposes the *Lebenswelt* in the practice of science. We rely on everyday observation of the world as it is "pregiven" to us in the practice of scientific experimentation. The manifest image is constantly used to give empirical content to scientific theories. So, I see Husserl's critique as a starting point for building a value-dependent realism, a realism that takes seriously the phenomena of the everyday life-world and the experience of suffering, and recognizes that human values are even more fundamental to descriptions of these phenomena than mathematical accounts.

Putnam's Internal Realism

Hilary Putnam (1987, p. 4) distinguishes views such as that of Sellars (scientific realism) from the view that there *really are* such things as tables and pink ice cubes (commonsense realism). In holding that scientific realism is an inadequate picture of the world, Putnam is in sympathy with Husserl. Putnam's more recent work has attempted to take seriously commonsense realism as well as the challenges of social constructivism. He calls his picture of the world "internal realism," a position that insists that realism is not incompatible with conceptual relativity (Putnam, 1987, p. 17). Putnam distinguishes two philosophical perspectives: externalist and internalist. The externalist perspective takes the world to consist of some fixed totality of mind-independent objects. There is exactly one true and complete description of this world, and truth involves a correspondence of some sort between the external objects and words for them. In order to be perfectly sure that a given description of the external world is true, one would have to have a "God's eye point of view." The internalist perspective, on the other hand, holds that the answer to the question, What objects constitute the world? makes sense only from *within* a theory. There may be more than one "true" description of the world. Because a God's eye point of view is not a point of view we can ever have, internal realism rejects a correspondence theory of truth in favor of a conception of truth as some sort of ideal coherence of our beliefs and experiences (Putnam, 1981, pp. 49-50). Internalism is not to be equated with the utter relativism that would result were we simply to construct a conceptual system. Internalism takes seriously experiential input to our knowledge; knowledge is constrained by more than coherence of beliefs. Internalism, however, holds that all knowledge is shaped to some extent by our conceptual and theoretical commitments. Therefore, there may be more than one true description, depending on conceptual choices, of any given reality. Truth depends to a large extent on coherence of experience and theory. But truth is not to be equated with rational acceptability. Rather, truth is an *idealization* of rational acceptability. We consider a statement to be true if it would be justified under "ideal epistemic conditions" (Putnam does recognize that these conditions, like "frictionless planes" in physics, can never be attained but only approximated to a high degree) (Putnam, 1981, pp. 54-55).

Putnam (1987, pp. 18-20) gives this simple example of internal realism and conceptual relativity. Consider a world with just three objects, $x1$, $x2$, and $x3$. Now, consider the question, How many objects are there in this world? We have said there are three. However, the answer really depends on one's conceptual system, in particular, on one's conception of an "object." Carnap, Putnam says, would regard this world as containing three objects, $x1$, $x2$, and $x3$. But there are other systems of logic (such as that of the Polish logician Lezniewski) that take mereological sums of particulars to be objects themselves. Under this conceptual scheme, the world has seven objects: $x1$, $x2$, $x3$, $x1+x2$, $x1+x3$, $x2+x3$, and $x1+x2+x3$. If one considers the "null object" to be part of every object, then the world has eight objects. So, the answer to the question of how many objects there are in the world depends on which conceptual scheme one is using. Within each of those schemes there is a definite answer. The answer is not purely a matter of convention, for some objects are given in the world. We cannot simply stipulate that there are none. However, to ask which objects are real is not a question that can be answered unless one specifies a particular conceptual framework that specifies what counts as an object.

V. TOWARDS A VALUE-DEPENDENT REALISM

I have suggested that the concept and classification of disease is best understood within a framework of value-dependent realism. Such a framework will take into account the insights of the social constructivists about the theory-ladenness of scientific observations and the essentially social nature of building a body of scientific knowledge. It will go a step beyond this in insisting that science is not only theory-laden but also value-laden. But at the same time it will maintain that the existence of a reality that is independent of our theorizing about it is at least an open question. Value-dependent realism, therefore, seeks to go beyond Putnam's internalism. Despite the insistence of the logical empiricists that externalist types of questions make no sense, those kinds of questions do continue to be intriguing.

In Putnam's internal realism, it can sometimes be difficult to see where the realism, in any strong sense of the word, is. If questions about external realism are nonsensical, then it is hard to see how we can maintain that the world offers resistance to our theorizing about it. A possible alternative to simply declaring external questions nonsensical is to distinguish between the many "internal realisms" or "local realisms" that make up the various possible descriptions of the world and the bigger, external, or "global" realism.[7]

The global realist can accept the internal realist's assertion that there may be several possible adequate descriptions of the world. Descriptions differ in degree and level of detail, and the detail of any describable fragment of the world is potentially very large, if not inexhaustible. A number of different classifications may be possible for any set of reasonably complex objects (Vision, 1988, pp. 76-77). This is what we see in Putnam's example of the world containing three, or seven, or eight objects, depending on the conception of object. The problem for the global realist lies in evaluating competing, empirically adequate, but incompatible descriptions. But this problem may be nothing more than an inability to completely describe any reasonably complex state of affairs. Drawings such as duck-rabbits, which seem to the same viewer at one time to be one thing and at another time something else, are an illustration of this problem. It might seem that the figure can be described as both a duck and a rabbit, but these descriptions are incompatible. The figure itself is neither a duck nor a rabbit. It is a figure that represents both a duck and a rabbit (Vision, 1988, pp. 78-79). The figure is real in the sense of the global realist. It is not socially constructed in the relevant sense. However, the figure becomes meaningful only through some theory (and value)-laden interpretation.

What Putnam's analysis points out is that any world, even a simple one, may admit of multiple adequate descriptions. This need not trouble the global realist, however, for even a global realist can admit that one must resort to theory-laden descriptions to make statements about the world. But this does not entail that there *is* no world apart from descriptions of it. That sort of constructivist claim amounts to saying that there is no ambiguous duck-rabbit, but only a duck when we see a duck, *or* a rabbit when we see a rabbit. However, we do find the assertion that there is an ambiguous duck-rabbit meaningful, and even interesting, even if we can only perceive one possible description of the figure at a time.

Students of the history of medicine sometimes find what seems to be an ongoing stream of conflicting philosophies. Today, allopathic medical practitioners compete for patients with osteopaths, chiropractors, herbalists, acupuncturists, and a host of other healers espousing sometimes similar and sometimes radically different views of the nature of disease and the way to relief. How do we assess their claims? We want our view of disease to reflect the reality of an individual's pain and suffering as much as we want it to reflect the research in the basic sciences that have made Western medicine so empirically successful in treating diseases in the twentieth century. Taking a globally realistic view might be a way to reconcile some of these apparently conflicting philosophies into a coherent whole. Value-dependent realism, in particular, will try to bring about this reconciliation by appeal to a set of universally recognized values as a starting point for our claims of knowledge about health and disease.

Hardly a physician would agree with the claim that Western medicine has simply constructed a framework of disease. Our conception of disease and our classification of diseases both seem to reflect a reality that offers resistance to our theorizing about it. People get sick, suffer, and eventually die from various diseases and injuries. We might describe sickness, suffering and death in different ways, but the reality of such human experience remains.

If we take seriously the arguments of the social constructivists, however, we come to the conclusion that our knowledge always reflects a particular worldview, which utilizes a particular array of concepts to explain the reality of illness. We employ these concepts to describe the world that is given to us. Illness is as real as disease, even if it is described within a different conceptual framework. Translating the framework of illness into the framework of disease is the major task of diagnosis in medicine. Explaining the limits and possibilities of this translation is my task in the following chapters.

Central to my thesis is the claim that both fact and value are necessary for understanding disease. What I hope to show in the next chapter is that facts are not as objective as the strict scientific realist might think, and not as subjective as the strict social constructivist might think.[8] Likewise, not all values are as subjective as the scientific realist supposes. Value-dependent realism bridges the fact-value gap in its recognition that values are necessary for the determination of what the facts are. In a framework adequate to explain the concept of disease, classification of disease, and diagnosis of diseases, values are as necessary as facts.

What I will propose is that no one "internal" description of the world is adequate to describe disease, nosology, and diagnosis. Various "internal realisms" will apply. In formulating a diagnosis, one will necessarily need recourse to value judgments in order to make choices about which descriptions one uses and which descriptions one rejects. If, however, we can agree that some foundational values ought to inform these choices, we have at least opened the possibility for a global realism about disease and diagnosis.

CHAPTER 3

FACT VS. VALUE

I. INTRODUCTION

In this chapter I will examine the relationship of fact and value. The value-dependent realism I am advocating holds that all facts in medicine, including facts about diagnosis, depend on values for their specification. However, I will argue that there are objective values. If this is so, we can maintain the possibility that medical facts, even though they are built upon values, reflect an objective reality. This will avoid the sort of relativism that results from the extreme social constructivist view that we simply invent our facts in order to produce a world that is socially useful.

I will first consider the nature of facts. Any position that claims to be a realist position must attend to the relation of facts and states of affairs in the world. In coming to realize the complexity of this relation of fact to the world, we will see how several of the arguments for the strict fact-value dichotomy fail. Next, I will consider the nature of values. The variety of types of values in medicine and diagnosis is not often appreciated. I will show how our factual knowledge in medicine is built from the interaction of empirical observation and certain values, which I will call *foundational values*. I will consider two important concepts in medical knowledge, need and function, and show how these two factors add a value dimension to medical facts. In short, I will argue that there are no medical facts without foundational values.

II. FACTS

Ambiguity in Usage of "Fact"

Ramon Lemos (1986, pp. 525-527) distinguishes four senses of fact in actual usage. The first sense of "fact" is the term's use to designate anything that exists independently of its being thought by anyone. In this sense, ordinary objects and events are themselves facts. To use Lemos's example, the Eiffel Tower is a fact in the first sense, whereas mermaids and square circles are not. Naïve, or commonsense, realism is presupposed in this sense. If there are no mind-independent entities, then there are no facts. While this sense of fact is sometimes used in common language, it is not generally what philosophers mean when they use the term "fact." If there are several possible adequate descriptions of the world, or several possible understandings of what constitutes an object, then this sense of fact will be equivocal.

The second sense of the term "fact" is its use to designate a state of affairs that obtains. In this sense, "fact" designates nothing in addition to the obtaining of the state of affairs; that is, facts do not constitute an additional metaphysical category. The Eiffel Tower's being in Paris is a fact in the second sense because that state of affairs obtains; the Eiffel Tower's being in London is not a fact because that state of affairs does not obtain.

The third sense of "fact" is the term's use to designate either the obtaining or the non-obtaining of a state of affairs. According to Lemos, this sense is less used than the second sense but is distinct from it. In the third sense, the existence of "negative facts" is admitted. This is useful if one wants to assert a fact about some possible state of affairs that does not obtain. The state of affairs of the Eiffel Tower's being in Paris obtains, and so it is a fact. The state of affairs of the Eiffel Tower's being in London does not obtain. But it is a fact that the Eiffel Tower is not in London. So, facts in the third sense can designate either states of affairs that obtain or states of affairs that do not obtain. However, this is still not to say that facts are a class metaphysically distinct from states of affairs, for they are still nothing more than the obtaining or non-obtaining of states of affairs.

The fourth sense of the term "fact" is the term's use to designate true propositions. Propositions are the content of statements, sentences, and beliefs. The truth or falsity of a statement, sentence or belief is determined by the truth or falsity of the proposition it states. Propositions are not identical to statements, sentences, or beliefs because the truth or falsity of propositions is not dependent upon their being stated, written or believed. The proposition that dinosaurs existed until sixty-five million years prior to the time that is now called the twentieth century was true even when there were no humans to assert the proposition. Thus, the fact (in the fourth sense) that the Eiffel Tower is in Paris amounts to nothing more than asserting the truth of the proposition expressed by the sentence "The Eiffel Tower is in Paris." Facts in the fourth sense are identical neither to real objects, nor to states of affairs, nor to the obtaining or non-obtaining of states of affairs. "Fact" in the fourth sense is an alternative to "true proposition." Facts in this sense are what true statements assert.

These four senses of "fact" might be put into two fundamental categories. The first three senses capture the idea that a fact is a state of affairs. The first sense of fact, any existing x, could be analyzed as the state of affairs consisting of the existence of x. The third sense of fact could likewise be analyzed as a state of affairs. The obtaining of state of affairs x is just x. The non-obtaining of state of affairs x is itself a state of affairs y. The first three senses of fact, then, are three instances of one conception of fact, which amounts to the second sense. This conception, "fact as state of affairs," is a feature of the world. The fact need not necessarily be taken as a metaphysical entity distinct from the state of affairs of which it is composed, although it might be.

In the fourth sense, a fact is not a state of affairs, but a true proposition about a state of affairs. In this conception, "fact as true proposition," facts are metaphysically distinct from states of affairs, and metaphysically identical to propositions. "Fact" in this conception is simply a linguistic alternative to "true proposition."

I would maintain that facts and states of affairs are not metaphysically identical, but are closely related and reciprocally imply one another. As Virgil Aldrich (1989, pp. 81-84) puts it, states of affairs can be photographed, but facts cannot. This characterization recognizes that facts, insofar as they are expressed, are to some extent related to our conceiving them. But there may still be states of affairs independent of our conceptions of those states of affairs as expressed in factual statements. That is, there may be facts of which we are at present unaware, but which nonetheless imply some true proposition. Facts can be discovered. But there is an element of invention in the process of discovery, for our discovery of a fact will depend on the theoretical and conceptual framework we choose to express the fact. I should also note that not all propositions express facts in this sense. An example is a proposition about certain types of values, such as "Scientific theories ought to be elegant in their formulation."

What exactly constitutes a fact depends upon metaphysics, epistemology, a theory of truth, and semantics. Insofar as facts are what make true statements true, a theory of truth is important in understanding what a fact is. Indeed, as Richard Kirkham (1992, p. 37) has shown, much of the confusion over the nature of fact and truth stems from the fact that different theorists have actually been involved in different kinds of projects. Some have inquired into the metaphysical nature of truth; others have attempted to justify the use of "truth"; still others have concentrated on the linguistic use of "truth." My primary interest here is in the meta-physical question about facts and the relation of fact to truth. In the next chapter, I will consider how facts partially constitute our understanding of the metaphysical nature of disease.

Realism, Facts, and Correspondence

The simplest type of theory of truth for any sort of realist is a correspondence theory, in which truth is a property of a proposition that corresponds to a state of affairs that obtains. Kirkham (1992, p. 119) divides these theories into two types: correspondence as congruence, and correspondence as correlation. Correspondence as congruence holds that every true proposition corresponds to a state of affairs that obtains. Correspondence as correlation denies that true propositions mirror states of affairs exactly, but holds that the correlation between true propositions and facts is a result of linguistic conventions. Truth, however, still resides in a correspondence between states of affairs and language conventions about them. These two theories are held by Bertrand Russell and J. L. Austin, respectively.

Russell's Correspondence as Congruence

Bertrand Russell takes a fact to be the kind of thing that makes a proposition true.

> If I say "It is raining," what I say is true in a certain condition of weather and is false in other conditions of weather. The condition of weather that makes my statement true (or false as the case may be), is what I should call a "fact." If I say, "Socrates is dead," my statement will be true owing to a certain physiological occurrence which happened in Athens long ago (Russell, 1985, p. 40).

A fact, then, is a state of affairs that obtains, and thus makes a proposition true. States of affairs in the world either obtain or do not obtain; they are not true or false. Likewise, facts are not true or false; there are no false facts. It is the obtaining (or not obtaining) of states of affairs that makes our propositions true (or false). Thus, there are at least two propositions for every fact, one true and one false.

Facts, for Russell, are part of the objective world; they are not created by our thought (except for facts about our psychological states, which have their origin in thought but then become part of the objective world). Facts are expressed by sentences, and not by single words.

> The first thing I want to emphasize is that the outer world—the world, so to speak, which knowledge is aiming at knowing—is not completely described by a lot of "particulars," but that you must also take account of these things that I call facts, which are the sort of things that you express by a sentence, and that these, just as much as particular chairs and tables, are part of the real world (Russell, 1985, pp. 41-42).

A fact, for Russell, is a structured complex. If we assert "Othello believes that Desdemona loves Cassio," our statement consists of a subject, Othello, and a complex object consisting of Desdemona, Cassio, and the relation of loving. The statement is a complex, which consists of the subject and the complex object knitted together by the relation of believing. The complex object is the fact corresponding to the belief. A belief is true when it corresponds to some such complex, and false when it does not (Russell, 1950, pp. 124-128). The congruence between fact and belief is what makes a true statement true.

Austin's Correspondence as Correlation

J. L. Austin, despite his belief that our language does not "mirror" features of the world in any precise way, still holds the view that facts are what make a statement true. For Austin (1979, pp. 121-122)[1], all statements involve two sets of conventions: the descriptive and the demonstrative. Descriptive conventions correlate words with *types* of situations, things, etc., in the world; they characterize. Demonstrative conventions correlate the words with *historic* situations, etc., in the world; they refer. Statements have both referring and describing functions. When the historic (i.e., particular) state of affairs to which a statement is correlated by its demonstrative conventions (its reference) is "sufficiently like those standard states of affairs" with which the sentence used in making the statement is correlated by the descriptive conventions, we say that the statement is true.

> When a statement is true, there is, *of course*, a state of affairs which makes it true and which is *toto mundo* distinct from the true statement about it: but equally of course, we can only describe that state of affairs

in words (either the same or, with luck, others). I can only describe the situation in which it is true to say that I am feeling sick by saying that it is one in which I am feeling sick (or experiencing sensations of nausea): yet between stating, however truly, that I am feeling sick and feeling sick there is a great gulf fixed (Austin, 1979, pp. 123-124).

One problem with this type of theory is that it may be impossible to state the exact conditions under which a proposition corresponds to a fact (Hamlyn, 1962, p. 205). When is a historic state of affairs which is correlated with demonstrative conventions "sufficiently like" a standard states of affairs which is correlated with descriptive conventions? Those problems aside, it is important for our purposes to realize that in Austin's view, as in Russell's, true propositions are true by virtue of their correspondence with facts. The truth of statements is dependent on the nature of the world.

This conception of a fact was opposed by P. F. Strawson. On Strawson's account, *pace* Russell and Austin, facts are not metaphysical entities. For Strawson (1950, pp. 129-156), facts are what propositions, when they are true, state; they are not what propositions are about.[2]

What "makes the statement" that the cat has mange "true," is not the cat, but the *condition* of the cat, *i.e.*, the fact that the cat has mange. The only plausible candidate for the position of what (in the world) makes the statement true is the fact it states; but the fact it states is not something in the world. It is not an object; not even (as some have supposed) a complex object consisting of one or more particular elements (constituents, parts) and a universal element (constituent, part). . . . Roughly: the thing, person, etc., referred to is the material correlate of the referring part of the statement; the quality or property the referent is said to "possess" is the *pseudo*-material correlate of its describing part; and the fact to which the statement "corresponds" is the *pseudo*-material correlate of the statement as a whole (Strawson, 1950, p. 135).

Strawson (1950, p. 143) accuses Austin of confusing two things: (1) the semantic conditions that must be satisfied for "*p* is true" to itself be true and (2) what is asserted by "*p* is true."

There is good reason to retain facts as metaphysical entities along with true statements, however. Facts can enter into causal relations in ways true statements cannot (Kirkham, 1992, p. 138). Physical states of affairs can cause a person to feel ill; true statements cannot. Austin's account of a fact allows us to see how some of the social constructivist insights are incorporated into our facts. It recognizes the contribution of linguistic conventions to our statements of fact. But it remains robustly realist in maintaining that states of affairs in the world are the basis of facts.

Facts and Meaning

Following the influential work of Gottlob Frege (1949), philosophers have usually taken the meaning of expressions to have two components: the sense (intension) and the reference (extension). The interesting question this raises is whether extension is sufficient for analyzing facts in medical science. Frederick Suppe, who defends scientific realism against the attacks of Kuhn and Feyerabend, thinks that reference is all that is needed to understand scientific facts.

Science, on Suppe's account (1973, pp. 197-212), seeks truth, but only what he calls empirical truth. Facts are what empirically true propositions assert about the world. It is the *extension* of a proposition, that is, the reference of its propositional content, that fixes its empirical truth value. For the social constructivists, on the other hand, a fact is fixed by both a proposition's extension and its intension. According to Suppe (1973, p. 211), the constructivist would have to hold that "The Morning Star is seen in the morning," "The Evening Star is seen in the morning," and "Venus is seen in the morning" all state different facts. This is so because "morning star," "evening star," and "Venus" have different intensional content. However, the extension of these terms is identical, and so they state the same empirical fact. However, there might be true metaphysical assertions about the world that are not empirically true and even Suppe admits this.

The question, then, is whether such true metaphysical assertions are necessary for an adequate understanding of medical facts. Does an adequate account of medical facts require the inclusion of intensional content? The strict scientific realist would say that is does not; illness is no more or less than the scientific explanation, i.e. pathophysiological state of affairs, that causes the subjective feeling of being ill. But if facts about disease require "true metaphysical and ontological assertions that are not empirically true" to explain their meaning, then empirical truth is not adequate. I hope to show that this is the case in the next chapter.

Furthermore, Suppe's use of "empirical" truth is too narrow to include much of what is properly considered empirical in medicine. Certainly such medical terms as hirsutism, lethargy, and even pain are properly considered within the empirical realm. Yet each of these terms has intensional content without which the term cannot be properly understood. Part of the reason why intensional content is necessary comes from a certain vagueness in the empirical terms of medicine. Another reason why intensional content is necessary is the fact that instantiations of many of the empirical characteristics common in medicine cannot always be adequately picked out by reference to the empirical terms alone. "Lethargy," for example, does not admit of all-or-none classification. Thus, it would be difficult to extensionally define the class of all lethargic people.

III. VALUES

When I speak of values, I am speaking of values in the broad sense. The realm of value in this book extends far beyond the moral values that all thoughtful people will recognize in the practice of medicine. In this section, I will discuss the nature of value and then delineate the different types of values that play a role in medicine and diagnosis. I will argue that many of the values in medicine and diagnosis are objective. If this is so, my insistence that facts depend on values need not lead to relativism or subjectivism.

The Nature of Value

"Value" is an ambiguous term. Z. Najder (1975, pp. 42-46) identifies three senses in which "value" is commonly used in both everyday speech and philosophical literature. The first sense is *quantitative*: value is what a thing is worth. The second sense is *attributive*: value is a thing or property to which valuableness is ascribed. Thus, states of affairs or facts may be values in the second sense. The third sense is *axiological*: value is the idea, criterion or principle that allows one to make evaluations. The first sense is becoming increasingly important in medicine as economic forces shape the health care delivery system. It is important in such contexts to determine which conditions are "worth" more than others are when making budgetary allocations. The second, attributive, sense is also important in medical discourse. In this sense, such things as health, ability to walk, and freedom from pain are important values. The 'third, axiological, sense of value is that of a conceptual principle that comes to play in any of our evaluations. This is the sense that will require the most philosophical work. The axiological sense of value must be made explicit in order to justify our claim that health, ability to walk, and freedom from pain *are* values in the second sense, and to appreciate what other states of affairs ought to count as values in medicine.

In trying to understand the *concept* of value, we must differentiate between value and mere personal preference. William Frankena (1980, p. 80) finds the term "value" troublesome because it tends to cover up this distinction. We sometimes talk about our personal preferences as our values. There is a difference, however, between a person's *thinking* something is good and that thing actually *being* good. Personal preferences do play an important role in our ideas about the value of health and disease, but I will argue that there are other objective values that ought to be recognized as values by any person, whether or not that person has a preference for them.

We do criticize others, saying that a person has bad values. This raises another ambiguity with the use of the term "value." "Value" may be used to connote either good or bad. We might say that values occupy a range from positive (good values) to negative (bad values). I will refer to negative values as "disvalues."

Recognizing Values

John Sadler (1997) has presented a heuristic scheme for identifying or recognizing values in medical and scientific discourse. Values manifest themselves in one of three dimensions. The first is the linguistic dimension. Values manifest themselves in the meanings of words and sentences. The manifestations may be in simple or "thin" terms such as *good, beautiful,* or *immoral.* They may also be in "thick" terms such as *elegant, morbid,* and *sublime,* whose meanings have "both purely descriptive and evaluative semantic content " (Sadler, 1997, p. 548).

The second dimension is the causal dimension. This refers to the particular way values manifest themselves. Values may be expressed as "value commitments." Our commitment to scientific theories that are simple, coherent and comprehensive is an example. Values may also be expressed as "value entailments." These entailments follow from our worldviews. If one is a reductive materialist, for example, certain values will follow from that worldview. In the causal dimension, values may be expressed in a third way: as "value consequences." This recognizes that particular theories or practices have effects that "can guide ongoing social actions and as such are worthy of praise or blame." The Human Genome Project, for example, will likely produce many ethical, legal and social concerns (Sadler, 1997, pp. 548-552).

The third dimension is the descriptive dimension. This is a typology of the sorts of things that are valued by people. Sadler (1997, pp. 552-554) lists five descriptive dimensions—aesthetic values, epistemic values, ethical values, ontological values, and pragmatic values—but he admits that there are others.

One of my major interests in this book is to unmask the values that are hidden in some of the "thick" terms used in diagnosis. These covert values have important influence on the way we treat people. There are two areas we must address before trying to articulate what sorts of values come into play in medical diagnosis, however. The first concerns theories of value. Value theory tries to explain what is ultimately good as well as what is instrumentally good. The second area is the issue of how we justify our theoretical account of value. I will consider these two areas in turn.

Theories of Value

The Good

Although value is not synonymous with good, the two concepts are closely related. What is good ought to be valued. A major concern of theories of value, then, is to explain what is good and the relation between different goods. There have been several types of approaches to the question of what is good. One way to classify theories of the good is according to whether they are monistic or pluralistic. Monistic theories hold that there is only one kind of good, whereas pluralistic theories hold that there is more than one kind of good and that the different kinds cannot be reduced to one particular good. Hedonistic theories are probably the most widespread type of monistic theory. Hedonistic theories hold that all good is pleasure, or satisfaction of desire, or some closely related sense of satisfaction. Hence, experiences are to be valued insofar as they produce pleasure. These theories go back at least to the ancient Epicureans, and have their clearest expression in the classic utilitarian theories of Jeremy Bentham (1973) and John Stuart Mill

(1973). There are monistic theories that take other goods as the ultimate values: Aquinas's beatific vision or Nietzsche's power, for example.

Pluralistic theories, on the other hand, hold that there are many different kinds of good. It may or may not be possible to rank these goods. One might, for instance, hold that pleasure and health are both goods, but that the good of health is not reducible to the pleasure that health brings. The theory might rank pleasure higher (or lower) than health, or it might hold the two values to be incommensurable.

Values

Values might be considered to be either intrinsic values or instrumental values. Instrumental values are what serve as a means to some further end that has intrinsic value. The intrinsic value of a state of affairs is determined completely by the state of affairs in question, taken completely in abstraction from consideration of the values of any other state of affairs that is not part of it (Lemos, 1995, pp. 34-35).

In medicine, the promotion of health, the treatment and prevention of disease, and the relief of suffering would certainly count as values. Whether health is an intrinsic value or only a value that is instrumental to some intrinsic value, such as pleasure or ability to carry out a chosen life plan, will have to remain an open question for now. I suspect that health is both an intrinsic value and also instrumental for other human endeavors. I will discuss the meaning of terms like health, disease and illness and the nature of their value-ladenness much more in the next chapter. My argument here does not depend on settling this question of ultimate value, but only presupposes that health is indeed a value, either intrinsic or instrumental.

Theories of Value Justification

In this section, I will examine the question of how some values can be objective. As it turns out, there are several possible routes. Value-dependent realism, then, might be built on different sorts of justifications for values.

Although the influence of logical positivism has faded in the philosophy of science, it remains stronger in moral philosophy, and finds its expression in the metaethical position referred to as non-cognitivism. This is one strain of moral anti-realism. For the positivists, to say something is good is merely to express one's approval for it, or to recommend it. R. M. Hare (1952, pp. 163-179), for example, holds a prescriptivist account of value judgments. For Hare, to claim that something is a value is only to prescribe, or recommend it. Meaningful talk about value for the non-cognitivist is restricted to talk about the meaning of terms. The cost of this type of theory is to lose any sense of objectivity about values.

Another strain of moral anti-realism includes "error theories," such as that of John Mackie (1977, p. 35). Mackie's theory is an ontological theory. For him, talk about values can and does go beyond talk about the meaning of the terms in the discourse. However, his claim is that there is no objectivity in moral discourse. Morality is simply a societal invention. Mackie's error theory holds that claims of moral objectivity do have cognitive content, but that all such claims are false.

If we are to maintain that some values are objective values, we must face the question of where the objectivity of these values resides. We generally desire the things we value. But do we value things because we desire them, or desire things because they are valuable? Ralph Barton Perry (1923, pp. 27-28) opts for the former. He considers value to be any object of interest, interest being what is characteristic of "instinct, desire, feeling, will and all their family of states, acts and attitudes." An object "acquires value when any interest, whatever it be, is taken in it" (Perry, 1923, pp. 115-116). Value, then, is a natural, or empirical, property. Value is a relational property of being an object of interest. I should note here that holding such a naturalist view of value does not commit one to saying that value is nothing more than subjective preference. I will explain how this is so in the next section.

W. D. Ross (1930, pp. 75-78), on the other hand, argues that this sort of relational view of good and value makes it hard to see how one might attribute intrinsic value to any object. If one is to maintain that some things are intrinsically valuable, then it seems more reasonable to say that we desire them because they are valuable. Value, then, is a non-natural property, which resides in objects independently of anyone's desire for them. Values are known not by empirical observation, but through intuition. For Ross, intuition is an intellectual process. For other non-naturalists such as Max Scheler (1973), value is intuited through the emotions.[3]

In value-dependent realism, values are foundational. If this philosophy is to be a true realism, then the foundational values must be objective values. My present task, then, will be to defend moral realism.[4] I will argue that a form of ethical naturalism, which we have already seen in the theory of Perry, and which can be traced back to Aristotle (1985), is a plausible way to defend objective values in medicine and diagnosis.

Justifications for Moral Realism

Moral realists maintain that there are facts about morality. That is, at least some moral claims are true. What, then, is it that makes moral claims true? There are three general possibilities (Sayre-McCord, 1988, pp. 14-22). One can be a subjectivist and a moral realist by holding that value judgments can be made only relative to the values, desires, and preferences of the one who is doing the judging. To say "x is good" is essentially to say "x is good for me." The problem with this sort of view, as Geoffrey Sayre-McCord (1988, p. 18) points out, is that the statement "x is good" will not have the same meaning when uttered by different people. While such statements do have a truth value, they have a truth value only relative to the inner states of the speaker. The trouble with this position is that value judgments have no objectivity, even if they are cognitively meaningful.

An alternate subjectivist view is the "ideal observer" type of theory, in which something is good if it would be judged as good by one who is in an ideal position to make such a judgment. The ideal observer, according to the influential account of Roderick Firth (1952, pp. 333-345), would be: (1) omniscient about non-ethical facts, (2) able to simultaneously visualize all facts and the consequences of all possible acts, (3) disinterested, (4) dispassionate, (5) consistent, and (6) normal in other respects. Ideal observer theories are still subjectivist in the sense that ideal moral judgments are made subjectively by the ideal observer. They avoid the objection that subjectivist theories do nothing but justify individual relativism. However, they run into an insuperable epistemological difficulty: it seems impossible for human beings to approach the conditions of an ideal observer and know what the ideal observer would decide.

The second possibility for the moral realist is to claim that what gives moral statements truth values is intersubjectivity. Truth about moral claims would rest on intersubjective agreement. This alone, however, quickly leads to little more than a system of cultural relativism. It is quite possible that a group of people might agree to adopt some ethical standard that is just wrong. One attempt to avoid out-and-out cultural relativism while maintaining that the truth conditions of morality arise from intersubjective agreement is a strategy parallel to the ideal observer theory. That is, the basis of morality is taken to be the intersubjective agreement among theoretical people under ideal conditions.[5] This, however, runs into the same epistemological problems that any ideal observer theory faces.

The third possibility for the moral realist is the objectivist position which holds that truth conditions of moral claims are dependent on neither the subjective states of individuals nor group practices, real or ideal. For the objectivist, values exist in themselves in some sense. Values might be taken as non-natural properties, as in the theories of Ross (1930) or Moore (1993). For Ross, "right" and "wrong" are terms that cannot be further analyzed. For Moore, these terms can be analyzed in terms of "good" and "bad," but the latter two are unanalyzable. On the other hand, an objectivist might be a naturalist, and hold that moral properties are analyzable in terms of non-moral properties.[6] In this way the moral good might be identified with some physical property or even a complex set of physical properties. Natural law theories fall into this category. However, to say that values are natural properties does not entail that they are physical properties. Perry's interest theory of value is an example.

The Constructivist Challenge
Ruth Anna Putnam does not defend a realist position, but argues that even if there are no objective moral values in the realist sense, there are objective moral values in another sense. She argues that we create both facts and values. Both are created in response to certain needs. Facts are constrained by our actual sensory inputs, by the conceptual framework embodied in our prior beliefs, and by a demand for coherence and consistency. Nonetheless, we create our theories, and we create our facts as warrants for our theories. Similarly, we create our moral values out of need for justifications for our actions. When we are confronted with a clash in values, our response is to create a new value. This creation is no more arbitrary than the creation of a fact, which is subject to the constraints mentioned above. We create our values out of already stable moral systems and attempt to reach a new reflective equilibrium (Putnam, 1985, pp. 187-204).[7]

We may create both fact and value in one sense, for what we take to be fact and value depends to some extent on our interests. However, we do not create fact and value *ex nihilo*. We build both fact and value out of the raw material that presents itself to us in our creative activity, even if we know the raw material only through our own worldviews.

The same sorts of considerations that inform the debate about realism in science are at work in the debate about moral realism. The value-dependent realism I am advocating recognizes a reality that exists independent of our theorizing, but a reality that is necessarily dependent upon some particular conceptual apparatus if it is to be described. Reality may allow more than one empirically adequate description of it. Likewise, value-dependent realism holds that there are objective values; all values are not mere subjective preferences. Again, however, there may be more than one way to describe them. Facts depend on the values that go into the construction and selection of the theories we use to describe them.

So, there are no facts without values. This is not to say that facts *are* values, but merely that there can be no judgments about facts without an interpretation that necessarily involves values.[8] Social constructivists and non-constructivists such as Karl Popper agree that there are no "bare" facts. That is, there are no facts that are not in some sense determined by the theory that underlies them. But values are embedded in scientific theory. As I argued in chapter 2, the theory we select to describe some bit of reality will depend on our worldview. This worldview will contain values, which I will enumerate later in this chapter. Although fact and value may be conceptually distinct, fact essentially depends on value.

Ethical Naturalism

A time-honored way to be a value realist is one that goes back to Aristotle (1985) and has its prime expression today in natural law theories: ethical naturalism. The naturalist holds that values can be derived from facts about the way things are. If this view is to be plausible, however, it needs to be further explicated. Nicholas Rescher shows that it is logically impossible to derive a value statement from purely factual premises without some enthymematic evaluative premise, but that such enthymematic premises are often available and amount to little more than truisms. For instance, the inference from

(1) Doing *A* would cause Smith needless pain.

to

(2) It would be wrong for anyone to do *A*.

requires only the addition of a premise such as "It is wrong for anyone to do anything that causes someone needless pain." And such a premise is really a truism. Anyone who would dissent from it, Rescher says, does not really know the meaning of morality. So, even though we must reject naturalism and its derivation of value from fact, we still have rational means to accept statements about value (Rescher, 1990, pp. 297-319).

It is not so clear, however, that Rescher's enthymemes will save his derivation of values from facts. First of all, Rescher's "truisms" do not follow obviously from the meaning of morality. If it is rational to hold that causing someone needless pain is always wrong, it is only because freedom from needless pain is an objective value. We first need to understand that freedom from pain is valuable before we can declare that it is wrong to cause someone needless pain. Thus, that it is wrong to cause needless pain is not the very meaning of morality, but one of the basic things that morality affirms once we do understand it.

Another defense of ethical naturalism is that of Alasdair MacIntyre, who cites A. N. Prior's counterexample to show how an "ought" statement can be derived from an "is" statement. From the premise "He is a sea-captain," we may validly infer "He ought to do whatever a sea-captain ought to do" (MacIntyre, 1984, p. 57). This may refute the strictly logical point Rescher is trying to make. However, it does not tell us anything substantive about the ethical obligations of sea-captains.

Hilary Putnam (1981, pp. 206-208) argues against G. E. Moore's assertion that goodness cannot be reduced to any physicalistic (natural) property. Putnam observes that Moore's argument depends on an implicit assumption that there is no such thing as a "synthetic identity of properties." According to Moore, goodness might be correlated with some natural property, but it could not be identical to that property. But if this is so, then Moore must also reject some accepted scientific discoveries, for example, the discovery that the magnitude temperature is the same as the magnitude mean molecular kinetic energy. Moore's argument would seem to suggest that temperature is a "non-natural" property, for the identity of temperature and mean molecular kinetic energy is not an analytic identity. One can know the meaning of temperature and not realize its identity with mean molecular kinetic energy. But, even though the concept "temperature" may be different from the concept "mean molecular kinetic energy," this does not mean that the corresponding properties are different.

Putnam takes Saul Kripke's notion of "metaphysically necessary" truths that have to be learned empirically to illustrate this point (Kripke, 1980, pp. 128-134). If we describe a possible world in which the fact that objects feel hot or cold is explained by a mechanism different from mean molecular kinetic energy, we do not say that in this possible world temperature is not mean molecular kinetic energy. Instead, we say that some mechanism other than temperature makes objects feel hot or cold. That temperature is mean molecular kinetic energy is necessary even though we cannot know it *a priori*. We have discovered the essential property of temperature empirically. What this means is that even though Moore may have shown that the word "good" is not *synonymous* with any natural property, this does not entail that the property of being good is not identical to some natural property (Putnam, 1981, pp. 207-208).

Supervenience

Another approach to showing that objective values may depend on natural properties, but not be *reducible* to natural properties is this: values might be supervenient on physical or social facts.[9] This notion of supervenience was explained by R. M. Hare in this way:

> Suppose that we say "St. Francis was a good man." It is logically impossible to say this and to maintain at the same time that there might have been another man placed in precisely the same circumstances as St. Francis, and who behaved in them in exactly the same way, but who differed from St. Francis in this respect only, that he was not a good man (Hare, 1952, p. 145).

Now, to be certain, Hare is no naturalist, thinking as he does that naturalism omits the crucial prescriptive force of moral judgments (Hare, 1952, p. 82). He is, in fact, a non-cognitivist. Hare does, however, think that supervenience is a peculiar characteristic of value-words. I will argue later that medical facts have frequent reference to biological functions (and malfunctions) and to needs of various sorts. Some values could plausibly be held to be supervenient on such facts. According to this account, these values would depend on natural facts, even if they are not identical with natural facts. In this sense, there is a fact-value distinction, but this kind of distinction is not the kind of fact-value distinction that plays a central role in Moore's "naturalistic fallacy." There still must be, on my account, some foundational values that go into making functions and needs facts of nature.

Types of Values

There are many different ways to classify values. These classifications encompass the previously discussed quantitative, attributive, and axiological senses of the term "value." I will here consider two approaches that will be helpful in making clearer the rather broad scope of my use of the term "value" in this book. The first approach is that of Nicholas Rescher, and considers the nature of values in terms of relations and purposes. The second is that of Ralph Barton Perry; it groups values in classes corresponding to different realms of human activity.

Nicholas Rescher suggests six classes of values, but admits that his list is not exhaustive.[10] It should also be noted that a particular value might fit more than one category.

First, values may be classified by subscribership to the value. This classification points out that values are values for someone. There are, for example, personal values and group values. An example of group values in the context of medical practice would be the professional values of doctors. The medical profession and its professional organizations set forth standards to which individual members of the profession are expected to conform.

Second, values may be classified by the objects at issue. These might be "thing values," desirable features of an object (purity of a gemstone) or an animal (speed of a racehorse), environmental values (beauty of landscape), individual values (intelligence), group values (mutual respect), or societal values (economic justice).

The third classification is a division by the nature of the benefit of the value. These sorts of values correspond to human needs and interests and include material and physical, economic, moral, social, political, aesthetic, religious, intellectual, professional, and sentimental values. Perry's approach toward "realms" of value is really a development of this type of classification. As I intimated earlier, however, these groupings are not mutually exclusive.

Fourth, values may be grouped according to the purposes they serve. Food value, medicinal values, and persuasive value are examples. These sorts of values come into play in discussions of need and function. I will consider these two concepts shortly.

The fifth classification is according to relationship between subscriber and beneficiary. Examples might be self-oriented and other-oriented values. The value of beneficent action in medicine is largely an other-oriented value, while professional values are largely self-oriented.

Finally, values may be classified by the relationship the value itself bears to other values. A distinction can be made between instrumental and intrinsic values. Certain values might be instrumentally valuable as mean to other intrinsic values. Another use of this classification would be to rank different intrinsic values in a pluralistic theory of the good, or to explain why the different intrinsic values are incommensurable.

For Ralph Barton Perry (1968, p. 14), the major realms of human life, those parts of human life that are universal or considered to be important, are specifically describable as "realms of value." I take it that these realms are roughly what Rescher has mind in his classification of values according to the nature of the benefit, but some of the realms would be more akin to Rescher's classification according to subscribership. A brief examination of these realms will show the expanse of the concept of value.

The first realm of value, and the one most readily recognized as a realm of value, is morality. Morality, in Perry's view (1968, p. 90), is the "endeavor to harmonize conflicting interests." Morality has to do with such things as what is good, what is right, and the nature of duty, responsibility, and virtue. Moral value plays an important part in medicine, and is an important component of what I will call *diagnostic* values.

A second realm of value is what I will call sociocultural value. This category encompasses several of Perry's categories, including social organization, social institutions, the cultural sciences, and conscience, which Perry (1968, pp. 137-200) takes to be a cultural system of approving and disapproving attitudes. We have already seen, in Fleck's account, the importance of sociocultural values in establishing what are considered to be scientific facts. Thus, sociocultural values are among the *foundational* values in diagnosis. We will see that sociocultural values, including professional values, play an important role in formulating the concept of disease as well as in the classification of diseases. Thus, sociocultural values are also among my *conceptual* and *nosological* values.

Political values form a third realm. Politics involves the formulation of a plan by which members of a society band together for mutual advantage, and to escape and resolve conflict. In addition, politics must create an instrument of enforcement to assure the reciprocity necessary to achieve the above goals (Perry, 1968, p. 202). Decisions about societal goals involve moral judgments, but they also involve a realm of value that is not strictly moral. We will see that political values come into play in decisions about research priorities. Such decisions have a direct impact on the establishment of new facts in medicine.

The law and jurisprudence is another distinct realm of value. Law is related to politics as well as to morality. One can make fairly straightforward judgments about whether or not a particular action conforms to law. The question of whether a particular law is good or bad, however, is another matter (Perry, 1968, p. 230). Such a judgment involves values that include, but also go beyond moral and political values.

Economic values deal with establishing the worth of things within a system of exchange in a society. This is a distinct realm of value, although, like legal values, it is related to morality and politics in important ways. Distribution of resources, wealth and profits fall into this realm (Perry, 1968, p. 248). Economic values come into play in the diagnostic process when deciding whether particular diagnostic tests are worth the cost.

Scientific values form another distinct realm of values. Science raises epistemological questions that are essentially normative. What sort of reasoning reliably leads to scientific knowledge? What is to count as proof? What is the metaphysical nature of facts in science? What is the value of scientific knowledge? Is such knowledge intrinsically valuable, or is it merely valuable for its technological applications? (Perry, 1968, pp. 307-336). Such questions are questions of value, not merely of fact, and are important foundational values in diagnosis.

Aesthetic values are widely recognized as constituting a distinct realm of value. What makes art distinctive is that it derives its character from a distinct set of interests. It is from these interests that aesthetic values arise. Flower gardens are works of art in a way that vegetable gardens are not because flower gardens are produced for the gratification of sensibility, as opposed to the practical interest of food production (Perry, 1968, p. 323). Aesthetic values are an important example of the conceptual values that go into producing the concept of disease.

The final realm of values I want to mention is that of religious values. One's religious beliefs can strongly influence one's moral values and even one's values in other realms. But religious values are a distinct realm, being derived directly from religious faith. Religious values play an important role in medicine. One's religious beliefs can exert a powerful influence on decisions about therapy. Jehovah's Witnesses refuse blood transfusions because of religious values. While these sorts of values do not usually exert a direct influence at the foundational, conceptual, or nosological levels of diagnosis, they can play a role at the diagnostic level by influencing what sorts of diagnostic tests are seen as permissible or required.

IV. THE FACT-VALUE DISTINCTION

The strict separation of fact and value has been widely advocated since the time of David Hume's famous dictum that one cannot derive an "ought" from an "is."

> But can there be any difficulty in proving, that vice and virtue are not matters of fact, whose existence we can infer by reason? Take any action allow'd to be vicious: Wilful murder, for instance. Examine it in all lights, and see if you can find that matter of fact, or real existence, which you call *vice*. In which-ever way you take it, you find only certain passions, motives, volitions and thoughts (Hume, 1978, p. 468 (Bk. 3, part 1, sec. 1)).

The strict fact-value separation is also at the heart of G. E. Moore's "naturalistic fallacy," that "good" in an unanalyzable property, and can never be equated with any sort of natural object (Moore, 1993, pp. 64-66).

Our consideration of the theory-ladenness of scientific facts, however, has shown that there may be several possible descriptions of any reality. We must make choices about which of these we will count as scientific facts. But whenever we make such choices, our values come into play in those choices. If this is so, our scientific facts are value-laden as well as theory-laden.

Arguments for the Fact-Value Distinction and Responses

There have been three kinds of philosophical defenses of the fact-value distinction: the argument from rational determinability, the argument from internalism, and the argument from ontological queerness (Railton, 1986a, p. 5). None of them, however, succeeds in dividing fact from value. Let us consider these defenses in turn.

Rational Determinability
Some defenders of the fact-value distinction hold that factual disputes can be resolved by appeal to reason and experience, while value disputes cannot. Two rational agents may agree on all the relevant facts and experiences in a given situation, and still disagree on the values involved. But, as Peter Railton (1986a, pp. 6-7) shows, facts cannot be distinguished from values according to what reason demands us to believe. Even deductive logic cannot tell us what it is rational to believe. If that were the case, then it would be rational to believe all the logical implications of our current beliefs, and this entails that we ought to believe an infinite number of things.

Railton's point is a good one. If we all subscribe simultaneously to several worldviews, which does seem to be the case, it is almost unavoidable that we should have some beliefs that are not logically consistent. For instance, a religious scientist may hold a materialistic view of a certain segment of nature while at the same time holding a non-materialistic view about God and the world as a whole. If one were to draw out all the logical implications from these two beliefs, there would certainly emerge a number of conflicting beliefs. Yet most of us would agree that it is rational for the scientist to maintain both the scientific view of a particular segment of the world, and the religious view of the world as a whole. Thus, we cannot depend solely on rationality to tell us what we ought to believe even in the realm of facts. Furthermore, value disputes can often be resolved through rational discourse by clarifying the nature of the dispute. Therefore, rational determinability does not neatly separate fact from value.

Internalism
According to the argument from internalism, to claim that I value something is a factual assertion. But to claim that something is in itself valuable is a different sort of claim; it is a value judgment, which a person

who does not share the value in question would not accept. According to internalism, the claim that something is valuable is something like a non-cognitivist commendation that the thing is worthy of being valued. If, along with this internalist view, we accept instrumentalism, the view that reason dictates only means to ends and not ends themselves, then we are left with a distinction between fact and value, for instrumentalism accepts no substantive ends (i.e., values) that all rational beings would have reason to pursue. Facts about values extend only to the realm of what *particular* people actually value (Railton, 1986a, pp. 7-8).

Railton challenges the internalist's claim that if something is to be intrinsically valuable for a particular person, it must be intrinsically valuable for all rational persons (a claim Railton calls "value absolutism"). If one rejects value absolutism in favor of "value relationalism," one can still maintain internalism and instrumentalism but reject the strict fact-value separation. There may be no such thing as absolute goodness, but only relational goodness: goodness for some particular being. A good might be intrinsic or merely instrumental toward other goods for a particular being. This relational conception of value treats value in a way similar to other relational terms. For example, nothing is absolutely heavy, but only heavy in relation to something else. Similarly, a relational conception of values "may yield an objectively determinate two-place predicate 'X is part of Y's good'" (Railton, 1986a, p. 11) and so avoid falling into the type of relativism that maintains that there are no objective values.

Railton makes a good point, but we may not even have to reject "value absolutism" to show why the internalism argument fails to drive a wedge between fact and value. Of course, relational goods are relative to the individual being considered, but they may still be considered objective in an important sense. Insulin is valuable to a person with diabetes mellitus; it may not be valuable in the same way to a person who is not diabetic. But the non-diabetic can still recognize the objective value of insulin. In this way values can be objective while still depending on the contingencies of the desires and needs of particular individuals.

Suppose, however, that we grant that values are only relational; that is, values are values only for particular people. It still seems plausible that some values will be universal simply by virtue of some characteristic that all humans share. If we take some very general description of our human (biological and otherwise) nature, we ought to be able to describe some general values that would be at least be a *prima facie* values for any human being. Potable water and nutritious food would be examples. But even some more specific values, such as being able to walk independently and without pain, would appear to be at least *prima facie* universal human values.

Queerness

The queerness argument comes from John Mackie and has metaphysical and epistemological components. Mackie thinks that there can be no objective values, that is, no facts about values. Such facts would be "entities or qualities or relations of a very strange sort, utterly different from anything else in the universe." To become aware of such entities we would need some special faculty of moral perception quite unlike those we use in our ordinary ways of knowing (Mackie, 1977, p. 38).

Even if a mysterious sort of moral intuition were necessary to intuit objective values, other sorts of realities, such as essence, identity, necessity, and possibility, also seem to require a mysterious sort of intuition if they are to be grasped. Indeed, even such a physical phenomenon as a photon exhibiting wave-particle duality in a double-slit experiment seem to defy our everyday grasp of what is to count as knowledge. Furthermore, we have seen that intuitionism is not necessary to justify a theory of value. So, I don't think that the queerness objection casts any more doubt on the facticity of values than it does on certain other philosophical and scientific entities.

The Importance of the Metaphysical Thesis

We have seen that some differences of opinion about the nature of fact and truth arise because people have been interested in different types of questions. The same thing has happened with the fact-value distinction. R. G. Swinburne (1961, pp. 301-307) has found three different types of theses about fact and value. The first concerns only the way we actually use language. On this account, the properties of descriptive (factual) and evaluative assertions are so different that we cannot directly infer one type of assertion from the other. The second type of thesis is normative; it concerns the way we *ought* to use language. The third type is the metaphysical thesis that there is a dichotomy between fact and value in the world.

As Swinburne points out, theses of the second type depend upon holding a thesis of the third type. Theses of the second type are themselves evaluative assertions. But if one wants to hold that there is a fact-value distinction, one cannot use purely factual assertions to establish the second thesis. One needs to introduce some evaluative premise into the argument in order to arrive at an evaluative conclusion.

As was the case with facts, I am here interested in the metaphysical thesis about the fact-value distinction. Metaphysical theses of the third type are crucial if one wants to say anything about the fact-value distinction beyond merely describing the way language is actually used (a thesis of the first type). However, the metaphysical thesis one holds about fact and value will depend on one's Weltanschauung (Swinburne, pp. 306-307) or, as I have put it, worldview.

When we ask questions about realism and social constructivism, we go beyond questions about the way language is used. We are asking questions about the metaphysical nature of entities. So, claims we make about the fact-value distinction in the context of our discussions about realism are claims of the third type, and, if Swinburne is right, presuppose some worldview.

This suggests an affinity with Kuhnian paradigms in science. For Kuhn, we cannot help doing science from within some paradigm. For Swinburne, we cannot help viewing the metaphysics of the fact-value question from within some *Weltanschauung*. Kuhn, however, does not mean to be doing metaphysics when he postulates scientific paradigms; he merely purports to be describing the way science is actually done. But science does presuppose a worldview, and in the received view at least, it is one that is materialist and free of value. If values play the role that I think they do in science and medicine, however, we need to reject the metaphysical idea that science is purely materialist, for values are not material. But to say that values are not material is not to say that values are not real, nor is it to deny that some values may depend in an important way on what is material.

Even those who hold a materialist view of science might admit that medicine has a component of human interaction that takes it beyond pure science. Thus, they might hold that medical science deals with pure facts, and values come into play only in the application of these facts in the context of medical practice. The position I am advocating, however, is that values enter into even the most basic level of scientific fact. It is clear that medical practice is value-laden, but even the basic medical sciences are value-laden.

The idea that values are merely subjective while facts are objective remains at the heart of scientistic thinking, the view that scientific method is the ultimate rationality and paragon of all right thinking. However, as Hilary Putnam has argued, "scientific" is not coextensive with "rational." There are whole domains of fact about which present-day science tells us nothing. One of those domains is that of objective values (Putnam, 1990, p. 143). Even those who would maintain that science is value free would admit that the practice of medicine involves value judgments. Such judgments can certainly be rational. Putting aside the strict fact-value distinction of Hume and Moore may allow a broader and more penetrating analysis of the medical diagnostic process than might be obtained from an analysis strictly in terms of science.

V. THE PLACE OF VALUES IN SCIENCE

Robert Veatch has argued that values come into play in three different stages of science and technology. In the "pre-scientific stage," scientists rank values in deciding what to study. Furthermore, the scientific enterprise itself depends on evaluating exactly what counts as rational discourse. Judgments about what constitutes objective procedure, precision, and accepted methods of verification all are value judgments (Veatch, 1976, pp. 126-131). In the "scientific stage," theory choice and hypothesis formation, choice of method of testing, choice of which data to observe and record, and selection of material to report all involve values (Veatch, 1976, pp. 136-140). In the "post-scientific or application stage," values will necessarily be involved and will differ depending on the sort of model in which the application of the science takes place (Veatch, 1976, pp. 143-149).[11]

In terms of medical science, it is in this third stage, the application of scientific discoveries, that most people would admit that value judgments play a role. This is the stage of making decisions about what sorts of treatments would be most desirable, and the like. In diagnosis, however, many value judgments at the pre-scientific and scientific stages are presupposed, but may not be immediately obvious to the diagnostician.

Diagnosis is set within the context of Western science and rationality, and this context is established at both the pre-scientific and the scientific stages. Questions about experimental procedure, levels of precision, and methods of verification, the most basic presuppositions of medical science, all require value judgments. The values involved are what I call *foundational* values. In chapter 2, in our consideration of social constructivism, we saw the influence of worldview on what we take to be scientific facts. We now can see what sorts of values make up the foundational values of diagnosis.

An important component of foundational values is the realm of scientific values. These values are involved in theory selection and selection of standards of measurement and precision. They are the source of the rules that stipulate what counts as acceptable scientific practice. Helen Longino (1990, p. 4) calls these "constitutive values," values such as truth, accuracy, simplicity, predictability, and breadth. Longino opposes these to "contextual values," which are "personal, social, and cultural values, those group or individual preferences about what ought to be." Sociocultural values certainly play a role in establishing what counts as acceptable scientific procedure. For example, the professional values of the relevant scientific community come into play in the incorporation of "journal science" into the realm of "vademecum science," to use Fleck's terms. Sociocultural and economic values are likely to influence what sort of research receives funding, and, therefore, what sorts of discoveries are admitted into the body of scientific knowledge.

An important component of foundational values are epistemic values. These values are part of the realm of scientific values, but also part of a broader sociocultural realm of values concerning what is to count as rational discourse. Marcello Pera (1994, p. 115) divides epistemic values into two classes: constitutive (distinct from Longino's constitutive values) and regulative. Pera's constitutive value is the acceptance of the agreement of cognitive claims with facts. This is a typical permanent value of the entire scientific tradition. Regulative values include "simplicity, economy, harmony, elegance, falsifiability, high degree of empirical content, fruitfulness, inter-theory consistence [*sic*], heuristic power, and so on" (Pera, 1994, p. 115). As Pera points out, participants in arguments about scientific merit must agree on interpretation of the values, whether a particular value is involved in the particular debate, and on the hierarchy of values, that is, which values are more important in the particular situation.

Also at the pre-scientific stage, values are involved in our decisions about the very concept of disease, that is, what constitutes the nature of disease. To study a particular disease requires at least some understanding of what constitutes that disease. The values involved in this issue are what I call *conceptual* values. I will discuss them in chapter 4.

At the scientific level, values will be involved in choosing the conceptual schemes that are used to classify diseases. This will involve choices about which data are significant and which data are not, and choices about the utility of grouping certain data together. The values involved here are what I call *nosological* values. I will discuss these in chapter 5.

The diagnostic process includes values at the pre-scientific and scientific stages as well as at the post-scientific stage. The values of the post-scientific stage involve what I call *diagnostic* values, which include not only the sorts of scientific values we have been considering, but also the moral values that come into play when decisions are made about subjecting patients to diagnostic procedures. Diagnosis, then, is not only the post-scientific application of science, but is itself a scientific practice.

VI. THE MEETING OF FACT AND VALUE IN MEDICINE

Two types of fact that play a prominent role in medicine include important foundational values. These are facts about functions and facts about needs. Disease is often understood as a defect in the function of a particular organ, gene, or physiological or biochemical process. Such functions and malfunctions appear at first glance to be purely a matter of fact. Likewise, facts about biological needs seem straightforward. It is a fact that human beings have needs for water, air and nutrition. These sorts of functions and needs, however, incorporate some important covert foundational values.

Functions

The modern period has been characterized by the acceptance of the fact-value dichotomy and the rejection of human teleology. Indeed, many have considered functional aspects of the human being as defined in terms of human nature not to be intelligible (MacIntyre, 1984, pp. 57-58). Along with Alasdair MacIntyre, some natural law theorists[12] have suggested that the way to bridge the fact-value gap is to recover a teleological framework for human life.

MacIntyre argues that functional concepts have a special character that makes it possible to derive evaluative conclusions from arguments containing only factual premises about functional concepts. If concepts such as "watch" or "farmer" are defined in terms of purpose or function, then it follows that the concept of a watch or a farmer cannot be understood apart from the concept of a good watch or a good farmer. Furthermore, the criteria for being a good watch or a good farmer can be specified in factual terms. A good watch is one that accurately keeps time, according to some specified tolerance of accuracy. A good farmer is one who, for example, gets a higher yield of crops per acre.

What MacIntyre suggests is that the alleged fact-value chasm depends on an assumption that no moral arguments involve functional concepts. This may be correct. However, his argument apparently amounts to a linguistic one. If a watch is defined as an instrument that performs the function of keeping time to a particular degree of accuracy, then the "goodness" of a watch is found not in an analysis of the function, but in the conformity of the watch to a definition that merely stipulates its proper function.

The problem is this: a watch may have several functions. Different people may value different of these functions. Some might find a stopwatch function essential for a watch to be a good watch, while others may be indifferent to this function.[13] There may be some function of a watch (e.g., keeping accurate time) that virtually everyone will value. However, what this shows is that people share values for particular functions, not that the watch is inherently valuable if it performs a stipulated function. Thus, MacIntyre's analysis does succeed in showing the intimate union of fact and value in function, but does not succeed in providing a way to derive values from facts.

A. W. Cragg (1976, p. 80) argues that functional descriptions entail negative evaluations. From the premises, "This auger will not bore holes," and "To bore holes is a necessary condition of being a good auger," we may validly conclude, "This is not a good auger." But functional descriptions also entail positive evaluations of at least a comparative type. Take the factual premises "A and B are augers," "A drills holes in wood," and "B does not drill holes in wood." Given the analytic premise "Any object which fulfills its function is better than an object of the same type which does not fulfill its function," we can derive the evaluative conclusion "A is a better auger than B" (Cragg, 1976, p. 80). One still needs to specify the conditions of being a good auger, however, and this will involve value judgments about the function: for instance, the kinds of holes one wants to drill, and the kind of materials through which the auger can drill. Thus, if one wants to specify facts about functions that go beyond a mere stipulation that something has a particular function, one needs to make value judgments.

Kenneth Schaffner shows how all explications of function have involved two components: a goal ascription and a causal claim. But choice of a goal is something extrinsic to functional explications. Goal choice cannot be reduced to causal language, and goals cannot be provided by dynamically construed evolutionary processes. In short, functional language cannot be reduced to nonfunctional language. Functional language in biology is heuristically valuable, however, because of strong analogies between organisms and subsystems and human artifacts (Schaffner, 1993, pp. 403-404). The success of Western medicine based on science is an eloquent demonstration of the heuristic value of functional language. The heuristic value, however, can tend to mask the fact that values are operative in any specification of human functions. We specify human function in terms of our choices of particular goals. Any particular human organ generally can perform different functions that lead to different goals. Goal choice depends on interests and values. The way we describe the function of the organ depends on the particular goal that interests us. So, facts about functions presuppose values.

Needs

Talk about needs is endemic in medicine. Physicians tell patients that they need an operation, they need to stop smoking, and that they need to take their medicine. Rollo Handy (1960, p. 158) suggests a definition of value that uses the concept of need: "x is a value = x satisfies a human need." The success of this proposal, however, hinges on how one understands "need." Handy (1960, p. 159) suggests that we leave the task to science to determine what is necessary for human functioning, especially with regard to biological functioning. Living organisms maintain themselves through complex homeostatic mechanisms. There are needs that must be filled if these mechanisms are to maintain the sort of life that an individual desires. Different sorts of life may give rise to different needs. But even given this difference, all human beings have a desire to satisfy what they perceive as needs.

However, it cannot be that science will tell us all there is to know about human needs. Some sociocultural values and aesthetic values, for instance, are outside the purview of science altogether. Despite the claims of sociobiologists and behavioral geneticists, it seems to me that science is hard pressed to explain many types of human social needs or the needs to experience beauty. As we will see in the next two chapters, some of these needs may fall within the scope of medicine by virtue of decisions of the profession about the domain of its expertise. This reflects the importance of professional values as well as sociocultural and aesthetic values in the determination of what counts as a need.

Paul Kurtz has distinguished between two types of needs: (1) biogenic needs, such as physico-chemical homeostatic needs for life maintenance and needs to grow, mature and develop, and (2) sociogenic needs, such as psychological security, needs for reciprocal love, to belong to a community, for self-respect, for creative expression, and for cognitive contact with one's own and external reality. Human beings all strive for these things. Value is connected to the "biogenic and sociogenic basis of the striving process" (Kurtz, 1958, pp. 555-568). We will see in the next chapter that the concepts of health and disease include social as well as biological elements. Hence "pure science" cannot hope to describe all human medical needs.

Donald Walhout also argues that the attempt to reduce all our experiential needs to biological needs is fundamentally wrongheaded. The existential incompleteness of the human person involves not only a need for biological cravings for food, air, etc., but also for needs that Walhout calls "spiritual."

> Thus need has reference to more than the biological instincts which must be satisfied in order to maintain the life process; it has reference to the whole of man's being. It might well be called ontological in character, involving incompleteness of being, existential deficiency, the presence of nonbeing with being. Thus the needs for knowledge, beauty, virtue, devotion, loyalty, and so on are just as properly needs as those economic and biological needs to which the word is usually confined (Walhout, 1978, p. 43).

Physicians often talk about medical needs as if they are pure facts: a patient needs a particular drug, or needs an operation. L. Duane Willard takes an opposing view: human needs are not facts about people, but are values. Needs are oriented toward goals, and goals are what people value.

"Need" is basically a motivational term which gives a sense of urgency to a desire. Disagreements about needs can be disagreements about facts, but they can also be disagreements about values (Willard, 1982, pp. 259-274). If this is the case, however, it would seem that needs are neither pure fact nor pure value, but include both fact and value. The human need for water is partly a biological fact, but it is also partly a value insofar as it reflects a desire to continuing living. In certain circumstances, such as when a person is in the final stages of a terminal illness, water may no longer be a need, not because the biological facts that make water a need have changed, but because the value of continued life has changed. So, alleged facts about medical needs are another example of entities that include both fact and value.

Needs, then, are analyzable in terms of particularly urgent desires, and this involves a judgment of value. Of course, we do not want to hold that needs are merely very strong desires. I might intensely desire to be a professional athlete, but if I do not have the requisite physical characteristics, no amount of training will allow me to satisfy my desire. In order to call a particular desire a need we need to make a further specification. John Rawls suggests that it is in the criticism of the ends desired that we can estimate the relative intensity of our desires. Rawls considers the formal end of human life to be the carrying out of a rational life plan. Desires may be irrational if the ends at which they aim are irrational. Desiring to live forever would be an example. Some desires may be inordinate, having been formed as a reaction to some deprivation. Desires may also be irrational if they are based on incorrect beliefs (Rawls, 1971, pp. 419-420). My desire to be a professional athlete would be irrational if I based that desire on my own deluded sense of my great athletic prowess.

I would like to suggest that the analysis of "need" in terms of desire or interest must primarily be understood in terms of rational desire or interest. This leaves ample room for individual preferences in formulating a rational life plan. However, at least a basic level of health is necessary for any rational life plan. This, and not merely a biological description of the human organism, is what makes it reasonable to speak of medical needs. It must be remembered that a plan of action suggested by a physician may not fit a particular patient's rational life plan. When a physician deems a treatment to be a need and a patient refuses it, it may be difficult to determine whether the refusal stems from a rational life plan, or from unfounded beliefs or fears. The physician ought to keep in mind, however, that medical needs cannot be determined from biology alone. Determining needs requires consideration of the values that go into the formulation of a rational life plan.

VII. CONCLUSIONS

Social constructivists have shown us that it is impossible to specify a "bare fact" outside any conceptual framework. But value realists have shown us that some of the values we take to be universal and important may have an objective basis. Value-dependent realism holds both of these elements to be important. Facts and values, although they may often be specified and referred to distinctly, are inseparable. The various types of values I have called foundational values are embedded in all the facts that will enter into the process of diagnosis in medicine.

PART TWO

DISEASE

CHAPTER 4

THE CONCEPT OF DISEASE

I. INTRODUCTION

Diagnosis is the process of giving a name to a particular condition, usually thought to be a disease. Before discussing the process of diagnosis, then, it will be useful to examine more closely what we mean by the term "disease." "Disease" can refer to the class of all diseases. It can also refer to various subsets of this class, such as pneumonia, and further subsets, such as bacterial pneumonia and viral pneumonia. Finally, "disease" can refer to a single instance of a particular subset of disease such as viral pneumonia. Furthermore, several terms such as "disease," "illness," and "sickness" are used in common parlance in an imprecise manner. In addition, virtually all disease classifications, which are called nosologies, include entities such as injuries, disabilities and deformities, which most people would not consider to be diseases. How we classify these conditions will be the topic of chapter 5. For now, my task is to examine the concept of disease itself.

This chapter will consist of two parts. In the first part I will consider the metaphysical nature of diseases. Do diseases such as bacterial pneumonia have some essence, or are they only names for similar phenomena seen in individual patients? I will argue that there are indeed constellations of pathophysiological and experiential phenomena that do tend to occur together, and are discoverable in the realist sense. However, our descriptions of these constellations are in an important sense socially constructed. We have a choice as to which clusters of phenomena we see as constituting a disease, and which we choose to interpret as background noise.

In the second part, I will examine the debate over whether the concept of disease is necessarily value-laden. Any description of physical findings in a sick person will already be imbued with foundational values. I have suggested that we choose which signs and symptoms are parts of a particular disease. We do so for particular reasons. This brings additional values into play. But the value-ladenness of the concept of disease goes even deeper. The very concept of disease seems to connote disvalue. In addition, disease is often associated with a failure of some part of the body to function properly. Function has been invoked to show that the concept of disease is value-neutral. However, I have already argued that the concept of function is value-laden. In this chapter I will show how the analysis of function brings further values into the concept of disease. All these values go beyond the foundational values that are bound up in the facts of sickness. This second level of values, those embedded in the concept of disease, I will call *conceptual* values.

II. THE METAPHYSICS OF DISEASE

One can take two extreme positions on the metaphysics of disease, roughly corresponding to the medieval notions of realism and nominalism. The disease realist holds that diseases are things in themselves. The disease nominalist holds that diseases are not entities in themselves. Rather, "disease" is only a name we apply to the state of individual people, each having a unique condition, albeit one that shares certain characteristics with the unique conditions of other people. Owsei Temkin (1961, p. 631) has used the terms "ontological" and "physiological" to describe these two general types. There have been many examples of these two conceptions of disease, although they have been called by several different names. I will use Temkin's terms as convenient labels and show how many different conceptions of disease fall into these two major groups.

Ontological Conceptions of Disease

Disease as Invading Entity
Roughly corresponding to the realist position is what Temkin calls the ontological conception of disease. A clear seventeenth century statement of this position is that of Thomas Sydenham (1624-1689). Sydenham (1979, p. 13) recognized that although diseases in different individuals may have different natures, they often resemble one another in having common symptoms. If all diseases could be reduced to definite *species*, Sydenham reasoned, then a rational mode of treatment specific for each disease could be devised. By "species," Sydenham and his contemporaries meant a "substantial form or essence," which exists in a Platonic realm and is manifest in the imperfect and transient objects of the physical world (Taylor, 1979, p. 8). For Sydenham, "plant" is a

species; he likened diseases to plants. "Thistle" is loosely applied to different sorts of plants, but the careful botanist observes differences in these plants and is able to sort them into different species. The careful physician aims to do the same thing with diseases by noting only the "clear and natural phenomena" associated with a particular disease, and ignoring the phenomena that are "accidental and adventitious" (Sydenham, 1979, pp. 13-14). Sydenham thought that nature is uniform and consistent in the production of disease, so that a particular disease would produce the same symptoms in all people. Diseases, in his account, are actually humors or miasmata that are drawn into the body and concentrate themselves into a species that becomes the disease entity.

> We must begin with noticing that humours may be retained in the body longer than is proper; Nature being unable to begin with their concoction, and to end with their expulsion. They may also contract a morbific disposition from the existing atmospheric constitution. Finally, they may act the part of poisons from the influence of some venomous contagion. From any one of these causes, or from any cause akin to them, the said humours become exalted into a *substantial form* or *species*; and these substantial forms or species manifest themselves in disorders coincident with their respective essences (Sydenham, 1979, pp. 18-19).

As Taylor (1979, p. 10) notes, this approach to nosography and diagnosis was at the time both novel and iconoclastic in that it rejected the doctrines of past medical authority in favor of objective observation. While Sydenham's methodological ideals and his view that specific diseases demand specific treatments continue to be influential, his particular metaphysical picture of disease as an invading entity raises difficulties not only for metaphysical physiologists but also for other ontologists. In particular, Sydenham's view of disease cannot explain the variations in manifestation of a particular disease in different individuals. If disease is an invading entity, it should manifest itself uniformly. If the individual body's response is invoked to explain the variation in manifestation of the disease, we have moved away from a pure ontological conception and toward a physiological conception of disease.

Disease as Internal Entity

The young Rudolf Virchow (1821-1902) explicitly repudiated the ontological conception of disease in favor of a physiological one. In his 1847 article, "Standpoints in Scientific Medicine," he writes:

> Ever since we recognized that diseases are neither self-subsistent, circumscribed, autonomous organisms, nor entities which have forced their way into the body, nor parasites rooted on it, but they represent only the course of physiological phenomena under altered conditions— ever since this time the goal of therapy has had to be the maintenance or the reestablishment of normal physiological conditions (Virchow, 1958, p. 26).

Nearly a half century later, however, Virchow came over to Sydenham's side in embracing his own ontological conception of disease, albeit one that was quite different from Sydenham's. In an 1895 *Festschrift*, "One Hundred Years of General Pathology," Virchow shows his commitment to cellular pathology, a way of looking at the body's structure in disease that was unknown in the time of Sydenham. However, Virchow now also commits himself to an ontological view of disease. He writes:

> In my view, the disease entity is an altered body-part, or, expressed in first principles, an altered cell or aggregate of cells, whether tissue or organ. In this sense I am a thoroughgoing ontologist, and I have always regarded it as a merit to have brought the old and essentially justifiable requirement that disease should be a living entity, and lead a parasitic existence, into harmony with genuine scientific knowledge (Virchow, 1958, p. 192).

Thus, Virchow came to regard the pathological cells within the body as the disease itself and not merely the cause of the disease.

Virchow's ontological conception of disease differs fundamentally from Sydenham's. Sydenham regarded a disease as a foreign parasite or miasma that invades the body from without. He saw diseases as fluid or humoral entities that entered the body and mixed with the body's own humors to produce characteristic sets of symptoms. Virchow's disease entities, on the other hand, were more or less solid and arose from alterations in the patient's own body cells (Taylor, 1979, p. 13). There is a sense in which Virchow's ontology of disease is more moderate than Sydenham's. For Sydenham, the species of a disease was metaphysically more "real," in a Platonic sense, than individual instantiations of it.

Virchow's disease was not so much an external entity as a particular cellular structure in an individual patient. Similar sorts of structures could, of course, be found in other individuals suffering from the same disease. So, for Virchow, diseases were ontological entities, not in the sense that they were independently existing invading entities, but because they were distinct parts of the people suffering from the diseases. The diseased part of the body was held to have a character of its own and in a strong sense to be self-contained (Taylor, 1979, pp. 14-16).

Physiological Conceptions of Disease

A different metaphysical conception of disease is known by various names including Hippocratic, nominalist, empirical, and naturalistic (Cohen, 1961, pp. 160-161). The major emphasis of this conception of disease is not on disease as a distinct entity, but rather on disease as a deviation from the normal. I will follow the lead of those authors who call this view the physiological conception of disease.[1] This name places emphasis on the functional and dynamic nature of disease, as opposed to the more static implications of the ontological view.

Disease as Humor Imbalance

The physiological conception of disease has roots in the humoral theory of disease that dominated medicine for about two thousand years, from the time of Hippocrates. Disease was not seen as an entity that invades from without, although invading entities might precipitate disease by upsetting the body's natural balance. Rather, disease was considered an upset in the body's natural balance of humors. This view is propounded by Plato in the *Timaeus*, where he takes the body to be comprised of the four elements of earth, fire, air, and water, and diseases to be caused by "unnatural excess or defect of these, or the change of any of them from its own natural place into another" (Plato, *Timaeus*, 82a, trans. Jowett, B.)

This physiological conception of disease is developed throughout the Hippocratic corpus by means of the humoral theory of disease. Disease is the imbalance of "powers" such as bitter yellow bile and pungent acidities. Relief of suffering comes only through "coction," the proper mixing of the humors (Phillips, 1973, p. 49). The writer of *Ancient Medicine* describes an upper respiratory infection. The nasal discharge is acrid and hot. But when the running becomes thicker and less acrid, the heat ceases. This is due to the discharge "being matured and more mixed than it was before." The writer says, "where acrid and unmixed humours come into play, I am confident that the cause is the same, and that restoration results from coction and mixture" (Hippocrates, *Ancient Medicine*, 18, trans. Jones)

In *The Sacred Disease*, the Hippocratic writer considers epilepsy, and says that this disease attacks the phlegmatic and not the bilious. The defect is an inability to purge the brain of excess phlegm, inhibiting the downward flow of air from the brain. Depending on where the phlegm is directed, it can cause palpitations of the heart, difficulty in breathing, diarrhea, or the classic signs of epilepsy (Hippocrates, *The Sacred Disease*, 8-16, trans. Jones).

As Owsei Temkin (1973, p. 398) observes, this conception of disease depends on a biologically conceived teleology. The body is so constructed as to allow functioning in a state of health, which is in accord with nature. When the humoral composition of the body becomes unbalanced, disease impedes the natural human functions. Disease is deviation from normal (the usual healthy state of the individual), and is not considered to be a metaphysical entity in itself, as it is in the ontological conception. An advantage of this view is that the physiological conception allows an understanding of how symptoms can be manifest at a distance from the seat of the disease. The symptoms of epilepsy, for example, occur far removed from the seat of disease in the brain.

Disease as Energy Imbalance
Although it arose in the context of the humoral theory of disease, the physiological conception of disease accorded well with the philosophy of the Enlightenment: health is the natural human state, which has been spoiled by the excesses and vices of civilization (Temkin, 1973, pp. 402-403). John Brown was an eighteenth century physician who maintained a physiological conception of disease in his *Elements of Medicine* (1780).

For Brown, the essence of being alive is the possibility of being affected by "exciting powers" in such a way as to produce phenomena that are peculiar to living beings. The exciting powers may be either external to the body or internal. The property of being able to respond to these powers is called "excitability" (Brown, 1804, p. 88-89, pars. 10-16).

Disease, on this account, is a function of excitement. When excitement is of a proper degree, the result is health. But when excitement is either excessive or deficient, the result is disease. Brown rejects the humoral notion of disease, but replaces it with a conception that is still a physiological conception. Instead of an imbalance of humors, disease is constituted by an imbalance of excitement: either too much (sthenic disease) or too little (asthenic disease).

> It is the excitement alone, through its varying degrees, that produces either health, disease, or recovery. It alone governs both universal and local diseases: neither of which ever arise from faults of the solids or fluids, but always either from increased or diminished excitement. Hence the cure is never to be directed to the state of the solids or fluids, but only to the diminution or the increase of excitement (Brown, 1804, p. 112, par. 62).

Disease is no foreign entity; it is not an entity at all. Exciting powers are necessary to maintain life. These same exciting powers, when in excess or deficiency, are what cause disease and even death.

> It is certain and indubitable that the exciting powers have one common effect. They produce the phaenomena peculiar to life—perception, motion, intellectual operation, and thinking. . . . Now since it is an universal law of nature that the same cause produces the same effect, it is evident that the mode of operation of the several powers above enumerated must be the same. . . . For, if there is no difference betwixt health and sthenic disease, except an excess of excitement in the latter, and none betwixt health and asthenic diseases, but deficient excitement in these last, what else can the operation of the remedies, in removing sthenic disease, be, but to diminish, and of those that remove the asthenic, but to increase the excitement? (Brown, 1804, pp. 279-280, par. 312).

This reduction of all disease to two simple types and all treatment to two possible types undercut all the nosologies that had been developed since the beginning of the Enlightenment. As Georges Canguilhem (1988, pp. 41-47) has insightfully observed, it is hard to understand why this theory of disease was so embraced by disparate practitioners of medicine throughout the Western world. What is interesting for our purposes, however, is not the specifics of the theory itself, but the metaphysical picture of disease that it gives us. It is thoroughly physiological in its view of disease as an imbalance of the same forces that are responsible for health.

Disease as Homeostatic Imbalance

The physiological view of disease stressed observation and experimentation rather than rational construction of nosological schemata. This was most clearly articulated by Claude Bernard in 1865. For Bernard, ultimate causes were outside the realm of science. The essence of things must remain unknown. All we can learn are relations, and reciprocal organic relations are what give rise to the phenomena we observe in biology and medicine.

> A salivary gland, for instance, exists only because it is in relation with the digestive system, and because its histological units are in certain relations one with the other and with the blood. Destroy these relations by isolating the units of the organism, one from another in thought, and the salivary gland simply ceases to be (Bernard, 1949, pp. 66-67).

In Bernard's physiological view, there can be no ontological conception of disease as in Virchow's conception of disease as an isolated malfunctioning organ. It is only in the "physico-chemical conditions of the inner environment" that we find the causation for the phenomena we observe. "The life of an organism is simply the resultant of all its inmost workings" (Bernard, 1949, p. 99). Attempting to articulate the essence of disease is, for Bernard, a waste of time.

> The words, life, death, health, disease, have no objective reality. . . . When a physiologist calls in vital force or life, he does not see it; he merely pronounces a word; only the vital phenomenon exists, with its material conditions; that is the one thing that he can study and know (Bernard, 1949, p. 67).

This is, roughly, the nominalist view that there are no abstract diseases in themselves, but only individual people with particular diseases. As H. Tristram Engelhardt, Jr. (1975, p. 131) points out, the physiologists wanted to show the dependence of diseases on general laws of physiology rather than particular laws of the pathology of particular diseases. Every individual disease state could be understood as a deviation from a general physiological norm. In this way the physiologists tried to avoid the philosophical problems brought about by trying to understand diseases as metaphysical entities. By avoiding this set of problems, however, they encounter another set of problems: how to determine what is to count as a physiological norm and what is proper functioning.

Disease as Temporal Construction

In the late nineteenth century, in the Polish journal *Krytyka Lekarska* (*Medical Critique*), a vigorous debate on the philosophy of medicine emerged. The authors took a generally holistic teleological approach to the philosophy of medical knowledge and discussed the influence of theory on medical observation and the classification of diseases (Löwy, 1990a). Even though many of its members were materialists, the Polish School recognized that medicine is more than a biological science.

Tytus Chałubiński is considered to have founded the Polish School of Philosophy of Medicine with the publication in 1874 of his book, *The Method of Finding Therapeutic Indications*.[2] Chałubiński's holistic approach led him to regard diseases as specific to individuals and to reject any sort of ontological conception of disease. He did not deny that specific organs and functions allow for a degree of uniformity of disease processes. However, he saw the laws of physiology and pathology to be so complex that they were able to bring forth a virtually unique disease in every individual. Diseases, then, are not ontological entities but "abstract representations of certain phenomena separated from their natural context of individual life" (Löwy, 1990b, p. 16). Chałubiński proposed that the concept of "disease entities" be replaced by "disease moments." Each of these moments would represent "a single pathological manifestation of a given patient at a given moment of his disease" (Löwy, 1990b, p. 16).

Disease as Spatial Construction

Michel Foucault favors a spatial rather than a temporal analysis of disease. He sees eighteenth century medicine, with its emphasis on pathological anatomy, to have made a spatial conception of disease primary (Foucault, 1975, pp. 3-4). The "birth of the clinic" in the late eighteenth century fundamentally reversed the ontological conception of disease so prevalent among the nosologists who worked in the early part of the eighteenth century and before. The clinic, for Foucault, is not merely the space where patients are treated, but is a discursive method of knowing disease. It is characterized most importantly by the ability to open the medical "gaze" at autopsy to the manifestation of diseases in the internal organs. The clinic is organized around the knowledge of disease gained from the autopsy.

Before the clinic, the "space of localization," the place where a disease is located, the site of inflammation and so on, was divorced from the "space of configuration," the place where a disease gets its "objectivity" by means of theoretical conceptualization (Foucault, 1975, pp. 3-4). The site of a disease might be far removed from its manifestations in the body. With the advent of nineteenth century anatomical observation, however, there was no longer need to conceive of a "pathological essence" beyond the symptoms of an illness. Disease became nothing more than the collection of symptoms, of which Foucault writes:

> They are nothing more than a truth wholly given to the gaze; their link and status do not refer to an essence, but indicate a natural totality that has only its principles of composition and its more or less regular forms of duration (Foucault, 1975, p. 91).

The clinical gaze has the effect of giving a nominalist reading to the concept of disease. "Diseases have no other reality than the order of their composition" (Foucault, 1975, p. 118).

The Metaphysics of Disease: An Assessment

We have been taking the debate between the ontologists and the physiologists to be a sort of reincarnation of the medieval debate between realists and nominalists. It is true that the ontological view has a strong realist motif, while the physiological view tends toward nominalism in its assertion that disease is not an invading "parasite," but rather the upset of normal constitution of the individual. However, as Engelhardt (1975, pp. 131-132) has shown, physiological theorists of disease need not have a commitment to nominalism. The physiologist (as well as the ontologist), however, does need some way to recognize the universality of some disease entities, if only for the pragmatic purposes of rendering effective treatments and making prognoses.

The physiologist need not reject the idea that disease is real. The realist nature of disease for the physiologist, however, will be conceptual rather than material. Diseases are not things in the same sense as rocks, trees, furniture and ice cubes. But the signs and symptoms, the things out of which we construct our diseases, are forced upon us by our experience in the world, even if they must be interpreted in order to call them signs and symptoms. As Lester King (1954, pp. 199-203) aptly put it, "Things are real, but are unthinkable except in some pattern. Patterns and relations are real, but are unthinkable apart from something which is related."

A major problem with the ontological conception of disease is that it tends to ignore etiology of disease, or at least it fails to adequately distinguish etiology from the manifestation of disease. Sydenham, as we have seen, thought that nature was uniform in producing symptoms for any given disease. He thus would have been less interested in searching for remote causes of disease than in simply cataloging observations of clusters of symptoms. A related difficulty for the ontologists is fitting individual variations in the manifestation of disease into ontological categories (Temkin, 1961, pp. 632-633).

The physiological approach, on the other hand, with its emphasis on imbalance, focuses on the individual and admits our contemporary concern with quantification and statistical deviations from normal. But as Temkin (1961, p. 637) observes, the physiological conception of disease runs headlong into the problem of where to draw the line between what is normal and what constitutes an imbalance. The physiologist is challenged to "show cause why the endless variations of form and gradations of function should somewhere admit classification as healthy and diseased." The physiological conception, in giving primacy to the individual, makes it difficult to draw boundaries between health and disease for collections of individuals, except by means of statistical deviations from population means.

Georges Canguilhem recognizes that the boundary between the normal and the pathological is not precise. But this does not imply a continuity between health and disease save for quantitative variations. Canguilhem sees the borderline between the normal and the pathological to be imprecise only when several individuals are considered simultaneously. When a particular individual is considered over a period of time, the borderline becomes evident to the individual (Canguilhem, 1991, p. 182). This way of seeing disease turns upside down the idea that disease is a deviation from normal functioning. The normal person knows normality only through a world where every person is not normal. For Canguilhem, it is disease that is the primary concept. The experience or the realization of the possibility of experience of the pathological is the basis for understanding the normal.

> In the case of disease the normal man is he who lives the assurance of being able to arrest within himself what in another would run its course. In order for the normal man to believe himself so, and call himself so, he needs not the foretaste of disease but its projected shadow (Canguilhem, 1991, p. 286).

Health and disease for Canguilhem, then, are qualitatively different states, and not just quantitative variations from a physiological norm.

Roberto Mordacci (1995, pp. 475-478) recognizes the multiplicity of meanings of the term "health." His approach is a phenomenological one, an analysis of the experience of health and disease. He plausibly argues that although there is an *experiential* priority of illness as a negative experience, there is a *logical* priority of health as a necessary condition for illness to be recognized as negative. Mordacci's account of health shares the spirit of Canguilhem's, however. Health is an "analogy of plenitude," an ability of the organism to successfully meet the challenges of the environment (Mordacci, 1995, pp. 487-490). Health is more than just the absence of disease. One advantage of this "positive" conception of health is that it allows us to analyze the concept of disease without *fully* explicating the notion of health.

This, and Bernard's suggestion that disease must be understood relationally, allows us to see disease as a conceptual apparatus that allows us to talk about, classify, and have some control over the constellations of signs and symptoms that we observe in sick people. It is, however, a conceptual apparatus that is necessarily bound to fact. That is, there are facts about people who experience illness, and these facts are the raw materials from which we build our conceptual diseases. Furthermore, empirical observation tells us that certain clusters of phenomena, which we have come to call diseases, have a relatively constant occurrence. These sorts of facts are necessarily value-laden in the sense that they carry the foundational values necessary to specify what counts as a fact in the domain of medicine. These foundational values are generally uncontroversial. They are the sorts of values that form the basis for our theories of empirical observation. However, we build our concepts of disease by observing the manifestations of what we call disease and constructing relations between facts; here there may be more disagreement about the values that come into play. So, for example, there is no disease without Virchow's structure of altered cells taken as an ontological entity. But there is also no disease without a recognition that this entity is related in a significant way, usually in some explanatory sense, to a constellation of signs and symptoms that we judge to cohere and to deviate from the norm we consider to be health. This interpretation of the metaphysics of disease thus incorporates elements of both ontological and physiological conceptions of disease, and can, therefore, serve as a bridge between them.[3]

There is a sense in which this way of thinking makes diseases metaphysically real. Diseases incorporate facts about pathophysiological states of affairs that may properly be considered as entities themselves. But diseases are not metaphysical entities until we have done a fair amount of observing and categorizing. Furthermore, we cannot call something a disease until we have judged what sorts of states are desirable and undesirable, that is, what are proper norms of health. Disease as a metaphysical entity is thus partially, but significantly, socially constructed. This element of construction opens the way for us to see how *conceptual* values enter into the concept of disease in a way that goes beyond the way foundational values enter into the basic facts of science out of which we build disease. In the next section, I will consider the role of conceptual values in the concept of disease and argue that this role is in fact a necessary role.

III. THE NORMATIVE NATURE OF DISEASE

My approach to disease is realistic insofar as it recognizes that disease is a concept built from facts concerning states of affairs naturally occurring in a body. These facts, however, depend on the foundational values that constitute the states of affairs as facts. Therefore, it should not be surprising that when we state facts about disease, we are also affirming values, even if they are covert. In this section, I will argue that disease itself is a value-laden concept. The concept of disease incorporates another level of values. Disease is a concept built from facts laden with foundational values and another level of values, which I will call *conceptual* values. When facts about diseases seem incontrovertible, it is only because all the values in them are accepted without question.

I will first consider the claims of some non-normativists who hold that the concept of disease can be analyzed without any reference to values. I will show how the method they use does indeed rely on values. I will then turn to the arguments that the concept of disease does depend on values. The claim that the concept of disease involves value judgments is known as normativism. I should at the outset distinguish between strong and weak normativism. Strong normativism holds that the concept of disease has *only* a normative component. A negative value judgment would be necessary and sufficient to call something a disease on this account. Weak normativism, on the other hand, allows a factual or descriptive component to enter into the concept along with an evaluative component (Boorse, 1975, pp. 51-52). I will defend only weak normativism.

The Non-Normativists

Boorse

Christopher Boorse distinguishes between disease and illness. He argues that disease can be specified without regard to values. Disease depends on species-specific natural functioning. Functions are attributed to parts of the organism, and the functions of the parts are causal contributors to empirically determinable goals. This natural functioning is understood as normality. If species have a natural design that can be empirically determined, then disease can be empirically determined, as disease is nothing more than a statistical deviation from normality (Boorse, 1975, pp. 56-58).

This account requires specification of the species, which serves as the reference class. It also requires specification of the class's normal function. Boorse takes normal function to be the statistically typical contributions to the survival and reproduction of individual members of the class. The basic notion of function on this account is of a contribution to a goal. For Boorse, once the goal is specified, function can be understood non-normatively. Disease is an internal state that reduces one or more of these functions below typical efficiency (Boorse, 1977, p. 555).

So, disease, in Boorse's account, is established empirically. Determining statistical deviations from normal functioning is taken to be a matter of empirical discovery, with no need to appeal to values. Illness, in contrast, does depend on values. Some, but not all, diseases are illnesses. A disease is an illness only if it is (1) undesirable for its bearer, (2) an entitlement to special treatment, and (3) a valid excuse for behavior that is generally unacceptable (Boorse, 1975, p. 61). The normative nature of illness is obvious in all these three criteria.

Boorse takes this account to explain the difference between diseases and undesirable conditions that are not diseases. For instance, hemophilia is a disease while inability to regenerate severed limbs is not. This is because blood clotting in wounds is typical of the species, while limb regeneration is not. So, even though the inability to regenerate limbs is an undesirable state, it is not a disease. The account also allows that asymptomatic conditions can be diseases even when they are undetected.

Boorse finds merit in this account of disease because it allows the attribution of disease to animals and plants irrespective of any human interests in them (Boorse, 1977, pp. 563-565). What it does not account for are universal or nearly universal diseases such as arteriosclerosis and dental caries (Boorse, 1977, pp. 566-567). These conditions would not be considered diseases because they do not statistically deviate from species-specific norms.

There is much to be said in rebuttal to Boorse's arguments, but I will defer that until after we consider the position of another non-normativist. After that, I will look at the responses of some normativists and give my own assessment.

Kass

For Leon Kass (1975, p. 21), health is a positive concept, and not just the absence of disease. He takes an essentially Aristotelian approach to health and disease. Health, unhealth, and all diseases exist only in particular living beings. Kass (1975, pp. 28-29) sees health as a *natural* norm, revealed as a "standard of bodily excellence or fitness, relative to each species and to some extent to individuals." Health is the working well of the individual as a whole. Disease entities may in some cases be constructs, but the symptoms and deviations from health that they group together are not (Kass, 1975, p. 23).

Kass thus agrees with Boorse about the possibility of specifying what counts as a state of health without regard to values. Like Boorse, he finds the ability to empirically discover an organism's normal functioning to be the foundation of determining whether an organism is healthy or not. Unlike Boorse, however, he attributes functions to the organism as a whole. This, I will argue later, is even more problematic for the non-normativist than Boorse's notion of normal function.

With regard to disease, however, Kass begins to move toward the kind of position I am advocating when he recognizes the social constructivist element present in disease entities while still holding a realist notion of the symptoms and deviations that make up the disease. But he does not go far enough, for his notion of deviation from an empirically determinable norm runs into the same difficulties as Boorse's position.

The Normativists

Lester King (1954, pp. 195-196) observed some years ago that the concepts of health and disease carry with them implicit values. First, health is good and desirable, while disease implies something bad. Second, statistics alone cannot label a deviation from normal as a disease. Someone with an extraordinarily high level of intelligence or athletic ability deviates from normal, but is not diseased. These two claims form the heart of the normativist concept of disease. I will here consider a few of the most important developments of this position.

Margolis

For Joseph Margolis (1976, p. 243), the distinction between illness and health depends not on the presence or absence of values, but on whether or not the condition is manifest to the bearer. A disease is either that which causes a "diseased state" or the diseased state itself. Illness, by contrast, is simply a diseased state made manifest to its bearer by means of symptoms. This, I think, is substantially correct. When I speak of illness in this book, I am referring to a patient's experience of feeling sick. One might have a disease, and not have the experience of having an illness.

This distinction is related to Alvan Feinstein's concept of a "lanthanic" patient, an asymptomatic patient who nonetheless is in a diseased state. The lanthanic patient's disease may be clinically evident, as in an asymptomatic person whose electrocardiogram shows changes pathognomonic of coronary heart disease. The lanthanic patient's disease may also be completely subclinical, as when a previously unsuspected pathological condition is found at autopsy (Feinstein, 1967, pp. 145-146).

This conception of disease presupposes some medical norm by which disease is differentiated from absence of disease. Boorse's candidate for this medical norm is normal functioning. But, as Margolis points out, a disease or incipient disease state may be present without causing any malfunctioning whatsoever. This is true provided one does not trivially define the incipient disease process as itself a form of malfunctioning (Margolis, 1976, p. 243). That would beg the question about whether or not function can serve as the norm for judging disease.

Margolis, like Boorse, makes use of the concept of function in his account of disease, but Margolis uses function to show the essential value-ladenness of the concept of disease. He criticizes theories of function such as that of Wright (which I will later consider in more detail). These theories rely on a distinction between the human animal (as member of a biological species) and the human person (as culturally trained in language, and thus capable of self-reference and cultural contribution). The trouble with defining human function in strictly biological terms is that there is no way to separate the functioning of the human animal from the functioning of the human person. We cannot adequately specify human function without reference to the goals of the human person. Therefore, because the function of the human person can only be defined in the context of cultural values, there is no way to specify human functions in value-free terms (Margolis, 1976, p. 251).

There is, however, a set of values, which Margolis calls "prudential values," that are "relatively neutral" to human beings' condition as animals or persons. These are such values as avoidance of death, prolongation of life, reducing pain, gratifying desires, and insuring security of person and property. Because the human body, unlike social institutions, has changed relatively little over the millennia, the biological functions of the body appear to be stable and empirically discoverable. But this should be explained by the conservatism of the prudential values and not by any appeal to natural design and function (Margolis, 1976, p. 251). Our biological functions appear to be natural only because certain of our values are "relatively neutral," that is, generally accepted by all. Margolis's prudential values, then, are important conceptual values in the concept of disease. They are part of the second level of values in diagnosis.

Engelhardt

H. Tristram Engelhardt, Jr. (1976, p. 257) proposes that there is an essential core of disease, albeit an "abstract and meager" one, and that some of the coincidence in human interests that Margolis finds is indeed essential. Engelhardt stipulates "illness" as a set of observables considered to be pathologically "distressful, displeasing, painful, or dysfunctional." A "disease state" is a set of observations and measurements, which has become causally associated with a particular illness, or has shown itself to be a good predictor of the illness. "Disease" is a cluster of pathophysiological (or psychopathological) generalizations that serve to explain the illness, and allow for prognosis and therapy. Engelhardt thus relies on two distinct realms, which he calls universes of discourse: the universe of observation and the universe of explanation (Engelhardt, 1976, pp. 257-258).

Illnesses and disease states, on Engelhardt's account (1976, p. 261), identify sets of phenomena associated with (1) loss of functions that we consider to be properly human; (2) pain that does not play a role in properly human functions; or (3) judgments about human disfigurement and deformity. All three of the above require value judgments about what are properly human functions, what counts as disfigurement, and the like.

Value judgments also function in disease explanation. We must choose between competing explanatory models: pathophysiology or demonic possession, for instance. Even within the model of modern science, we choose explanatory models according to the types of foundational values discussed in the last chapter. In addition, we make conceptual value judgments in deciding which states of affairs to count as diseases for the social purposes of ascribing blame or blamelessness, and for putting certain individuals in the "sick role" (Engelhardt, 1976, p. 262).

Non-normative accounts rely on value-free descriptions of natural function. The problem with such accounts is that they wrongly assume that there is a standard environment in which we can empirically observe what is to count as proper natural function. Engelhardt argues that there simply is no single excellence or function for all individuals of the species. Furthermore, depending on species survival as a criterion for judging what functions are proper to a species ignores the emphasis we rightly place on the individual plight of the victim of disease and illness. Engelhardt thus rejects the claim of Boorse and Kass that disease reflects some natural norm. Nature, he says, "does not care for excellence, nor is it concerned for the fate of individuals qua individuals" (Engelhardt, 1976, pp. 264).

Engelhardt does seem to make one exception: the state of affairs that enables our activities as free, rational agents. Any definition of disease or assessment of function presupposes rational agency. This is the one characteristic, according to Engelhardt, that is typical of all persons as wholes (Engelhardt, 1976, p. 266). It does seem, however, that any assessment of what it is to be free and rational will involve making value judgments. Rational individuals differ even on what it means to be a free and rational person. This often becomes evident in difficult cases of trying to judge whether or not an individual is competent to make decisions about medical treatment. Value judgments would appear to be as operative here as in any assessment of health or illness.

Whitbeck

Caroline Whitbeck, like Leon Kass, holds a positive conception of health, that health is more than just the absence of disease. Medicine, in her account, actually has little to do with health, but concerns itself with diseases, injuries, and impairments. A disease is a psychophysiological process having the following characteristics: (1) people wish *to be able to* prevent or terminate the process because it interferes with the bearer's ability to do things that people wish and expect to be able to do; (2) either the process is statistically abnormal in those at risk or people have some other basis for a reasonable hope of finding means for prevention or treatment of the process; (3) the process is not also *necessary* for doing anything people commonly want or expect to be able to do (Whitbeck, 1978, pp. 210-211).[4]

According to this definition, disease is value-laden inasmuch as those things that people wish and expect to be able to do are specified in particular social contexts. Hence, sociocultural values will enter into what is seen as disease at the conceptual level. According to clause (1), diseases are the sorts of things that people *wish to be able to prevent or terminate*. This is a value judgment that may well vary from society to society and from time to time, as well as at different times in the life of a particular individual. The second disjunct in clause (2) is meant to include as diseases those conditions that may be statistically normal for a population, but still disvalued. An example is arteriosclerosis in the aged. Statistical deviance is not necessary to classify some condition as a disease if we reasonably come to think that we might be able to prevent or treat the condition.

Whitbeck recognizes three ways in which concepts may be value-laden. First is a "weak sense" in which all distinctions made in a language reflect what the community using the language takes to be important. The values involved in this sense are an important part of my foundational values. The second "strong sense" ascribes goodness or badness. In the case of the concept of disease, the strong sense attributes badness. The badness might be aesthetic, ethical, religious, prudential, epistemic, or hedonistic. Whitbeck takes the concept of health to be value-laden in this strong sense. The third sense is called the "capability sense." This, according to Whitbeck, is the interesting sense of the value-ladenness of the concept of disease. A concept is value-laden in this third sense if and only if "the concept or the definition of the term warrants the conclusion that people have an interest in *being able* to influence things of that type"

(Whitbeck, 1981a, pp. 614-615). This third sense of value-ladenness gives Whitbeck some attractive explanatory power that is lacking in the notion that disease is merely something that is disvalued. For Whitbeck, it is not necessary that a particular instance of disease be a misfortune for its bearer, but only that it be the kind of pathophysiological process that one wishes *to be able to* prevent or effectively treat. On this account, cowpox before the development of smallpox vaccine could still be considered a disease even though it was in fact valued as prophylaxis for smallpox. Likewise, sterility can be seen as an impairment even though some people go to trouble and expense to acquire it through medical intervention (Whitbeck, 1981a, pp. 614-615). So, on Whitbeck's account, diseases may not necessarily be disvalued by their bearers. Nevertheless, diseases are essentially value-laden in the sense that they are the sorts of phenomena that most people would like to be able to control.

The Value-Ladenness of Disease: An Assessment

Conceptual Values and Objectivity
The sorts of values that I call conceptual values include both Whitbeck's strong sense and capability sense. That is, I agree that a value judgment is involved in saying that a particular condition is the kind of condition that people want to be able to prevent or effectively treat. Thus, I see aesthetic, hedonistic, physical, sociocultural and religious values, among others, as playing important roles in decisions about what sorts of conditions we would like to be able to prevent or treat. While foundational values are largely covert in the fundamental concepts and methods of science and medicine, and only implicitly held by scientists and physicians, conceptual values are more likely to reflect overt individual preferences about personal goals and visions of the good life. This is not to say that all these are purely subjective preferences, however. Even if we cannot agree on a single ultimate goal of human life, some sorts of life goals are better than others by objective standards with which virtually any thinking person would agree.

James Lennox (1995, pp. 499-511) recognizes that the concepts of health and disease are value-laden. He argues that there are subsets of values that are formed in recognition of some fundamental facts about the biological organism. These are not moral values, but are normative nonetheless. Living things face a fundamental alternative: life or death. Medicine identifies life as a fundamental goal to be achieved by sustaining successful biological functioning. "Successful" is a value term, and not just a description of species-specific functioning. Success is measured objectively, however against the standard of maintaining life. In this way, the value dimension of disease can be recognized without sacrificing objectivity.

This is a beginning, but it does not go far enough. As Robert Sade (1995, pp. 518-519) makes explicit, biological survival is not the same as life. Successful living for the human being is not just biological survival; it is living a flourishing human life. Sade argues that a proper understanding of "flourishing" requires both impersonal, biological values and personal, moral values. The standard for judgments about flourishing, however, comes from "objective information about structure and function, supplied by the science of medicine." The objectivity of the values involved in health and disease is "not determined by science; rather, it is determined by the rational use of objective information supplied by science" (Sade, 1995, p. 523). There is much room for debate about what constitutes a flourishing life. Flourishing will, in fact, be different for different people. But this is not to say that we cannot make objective determinations about whether or not an individual is flourishing.

Kateryna Fedoryka (1997, p. 149) finds a difficulty with this type of a suggestion, and with all theories of health and disease based on value: the health of particular patients is defined in relation to their own interests. She has proposed a framework that tries to get around this problem while retaining a value-dimension in the concept of health. Well-being has to do with satisfaction of needs and desires. Fedoryka separates well-being that is a direct function of nature from well-being that is a function of individual choice. Health is the "dimension of the individual in which a certain unfolding of being happens in virtue of the natural structure of that being" (Fedoryka, 1997, pp. 150-155). A merit of the framework is that it is set on a foundation of value realism, and does not ignore the value dimension of health and disease in the way that some other objectivist theories do. It also avoids the specific problems that arise from taking the setting of goals to be a purely subjective enterprise. However, Fedoryka's appeal to the natural structure of organisms runs into the same difficulties we found in Boorse's theory: the problem of specifying the nature of organisms. It is here that the value component plays a major role and must be further specified.

Natural Functions

The non-normativist position relies on a value-free notion of natural function. The normativist holds that diseases are disvalued in themselves, or at least are the sorts of things that most people want to be able to control. The desire to be able to control a condition implies that it would be disvalued under some circumstances. I first want to show how values are inherent in the concept of function. Then I will discuss some of the complexities in the thesis that diseases are sorts of things that are disvalued.

In order to specify what is to count as a disease, non-normativists such as Boorse and Kass must rely on purely descriptive facts. Natural functioning constitutes normality and disease is taken to be a statistical deviation from some natural functioning. Boorse holds that natural functions can be empirically discovered. His account of disease, then, must specify some standard of normality, which appropriately reflects proper functioning with respect to the design of the organ in question. Boorse must conceive the organism, with its constitutive organs, as a sort of mechanical object (his example is a Volkswagen), which has an intelligible design and a discoverable function. Normality, then, is simply function in accordance with design. This is a mechanistic model of disease akin to the biomedical model discussed in chapter 1.

The trouble with trying to discover natural function for the human being, or even parts of the human being, is that we have no plans from the designer, as we do with Volkswagens. Thus, we have no standard from which we can ascertain whether the part is functioning properly. Because of this, Boorse's attempt to specify function as a causal contribution to empirically identifiable and value-independent goals must ultimately fail. As William Bechtel (1985, pp. 142-144) has recognized, Boorse's position leads him into a vicious circle. The notion of normality itself depends on the same notion of function it tries to explicate.

Bechtel suggests that evolutionary theory provides the basis for the discovery of a natural human teleology. Proper human functioning is whatever contributes favorably toward species survival (Bechtel, 1985, pp. 149-155). To make this work, Bechtel relies on Larry Wright's analysis of function. For Wright, to say that the function of X is Z is to say that "X is there *because* it does Z," or "Doing Z is the *reason* X is there," or "that X does Z is *why* X is there." "Because," "reason," and "why" have etiological force in Wright's account. Thus, function is essentially explanatory, and not merely evidential (as in "It is hot because it is red.") (Wright, 1973, pp. 156-157). However, Wright gives this example of why this etiological notion of function must be further qualified. The etiological reason for oxygen's presence in the blood is its capacity to readily combine with hemoglobin. But the function of oxygen is not to combine with hemoglobin; that is merely a means to the end of oxygen's function in necessary chemical reactions. If the atmosphere were suddenly to fill with carbon monoxide, hemoglobin would bind carbon monoxide. Carbon monoxide's affinity to hemoglobin is the etiological reason for its presence in the blood, but it would not be right to say that carbon monoxide's function is to bind with hemoglobin. Wright therefore specifies that the etiologies in functional explanations must be functional etiologies. This requires the addition of another condition to the above definition. So, "the function of X is Z" *means*

(1) X is there because it does Z,
(2) Z is a consequence (or result) of X's being there (Wright, 1973, p. 161).

What this account does is to explain current states of affairs in terms of their consequences. Bechtel introduces evolutionary theory into this scheme to get the following conclusion. The occurrence of the present X results from the fact that a past X did Z, which made X more adapted and so more likely to reproduce. Hence, something serves a particular function if it was selected in the past and now exists because it met an evolutionary need (Bechtel, 1985, p. 150).

Bechtel takes seriously Margolis's criticism of Wright: that Wright's account has the relationship between teleology and function backwards. One could imagine circumstances, says Margolis, in which the survival of the species depended on the combining of carbon monoxide with hemoglobin. This shows the difficulty of maintaining the asymmetry of the concept of "consequence" (Margolis, 1976, p. 250).

Suppose we were to reverse the order of connection between selection and teleology. As Bechtel recognizes, this would require us to find an independent way of determining what is functional. His proposal is that we define something as functional if it "increases the propensity of its bearer to reproduce." This reverses the direction of selection and function and so, he says, answers Margolis's objection (Bechtel, 1985, pp. 150-151). Why, however, should we be content with this notion of evolutionary advantage as the measure of proper human function? I have already mentioned Engelhardt's objection that there is no standard environment from which to judge what functions give selectional advantage. In addition, we might ask why we should regard species survival to be the key to understanding function. Selecting this as the key to understanding function is a value judgment. It is to select a theory that attempts to explain the biological diversity we observe and to claim that this *theory* is a value-independent observation. But this cannot be the case for any theory, for all the reasons I have given in chapter 2.

Furthermore, judging what functions have survival value cannot be a matter of simple empirical observation, for survival advantage of particular characteristics in the human species would take generations to ascertain. While such observations were being made, it is highly likely that major conceptual changes in medicine and disease will have occurred. Such changes may fundamentally alter views about the very entities being studied. One need only notice the conceptual changes that have occurred in the last century in such fields as immunology and genetics. These conceptual changes have radically altered the way we approach many diseases. Thus, several crucial background conditions are unlikely to retain the stability over time that would be necessary to make any valid determinations about human functions in terms of survival advantage.

Kass's conception of normality moves away from the mechanistic model of Boorse in that it takes health, or normality, to be the well-working of the organism as a whole. This requires that we discover some sort of Aristotelian essence that constitutes human well-working. But if we have no design plans for parts of organisms and thus cannot empirically specify functions for parts of organisms, how can we expect to empirically discover proper functions of entire organisms, especially ones as complex as human beings? Whether a human being is working well will depend upon the goals of the particular person. If anything is characteristic of human beings, it is the capacity to set their own goals. These goals are extremely varied and often original. They will demand extraordinarily varied physical states. Health for a philosopher may not be the same as health for an Olympic figure skater.[5] As René Dubos has remarked, the executive, the lumberjack, the monk, and the supersonic pilot have various sorts of physical and mental needs. What is a physical impairment for one might not even be noticed by another. Their goals determine the kind of "vigour and resistance" that is required for success in their lives (Dubos, 1968, pp. 67-68). Without the stipulation of proper function by means of an explicit design plan, any decision about whether something is functioning well or not will depend as much on individual expectations as on inferred natural design. These expectations are contingent on value judgments about what kind of life is desirable.

But suppose that we actually could discover a design plan for the human species. We still could not say that functional deviations from the design constitute disease, for some deviations might actually be improvements on the original design. Deciding which deviations were improvements and which were not would require specification of goals that go beyond the design plan. As Martin Bunzl (1980, p. 118) asks, "How could we hope to show that the highest-level 'goals' of the system were *its* goals, non-normatively, without appealing to some larger system in which the system played some role in contributing to that larger system's goals?" This suggests that, inasmuch as individual people are part of a larger society, sociocultural values will determine what an individual sees as normal functioning for the individual as well as for the society as a whole. These sociocultural values aim at myriad goals that cannot be reduced to Boorse's empirical goal of species survival. Boorse's suggestion that function be taken as a causal contribution to a goal actually pursued by the organism neglects the essential social and freedom-oriented goals of the human being. Medicine concerns *human* disease, and the language of disease ought to reflect those aspects of human life. Considered in this way, disease will range from "identification of syndromes, concurrences of undesirable symptoms based only on aesthetic, algesic or teleological criteria, to 'full' disease explanation involving causal accounts in terms of nomological structures" (Agich, 1983, p. 37).

The human being is immersed in an environment that constantly presents challenges to health, challenges that are constantly evolving. The human body responds to new environmental challenges by altering itself and establishing new norms. One might say that the body re-creates itself in response to those challenges. Canguilhem recognizes

> the fundamental biological fact that life does not recognize reversibility. But if life does not admit of reestablishments, it does admit of repairs which are really physiological innovations. The more or less large reduction of these innovation possibilities is a measure of the seriousness of the disease. As far as health in the absolute sense is concerned, it is nothing other than the initial boundless capacity to institute new biological norms (Canguilhem, 1991, p. 196).

Health, for Canguilhem (1991, pp. 198-199), is "a set of securities and assurances," a "biological luxury" that allows us to recover from our sicknesses. Disease is a "reduction in the margin of tolerance for the environment's inconstancies. Canguilhem realizes that biological functions are not static, but are reformed in response to challenges of the environment.

As Hans Jonas insists, survival alone cannot be an adequate evaluation of life. The goal toward which biological functions aim must be something more.

> If mere assurance of permanence were the point that mattered, life should not have started out in the first place. It is essentially precarious and corruptible being, an adventure in mortality, and in no possible form as assured of enduring as an inorganic body can be. Not duration as such, but "duration of what?" is the question. That is to say that such "means of survival as perception and emotion are never to be judged as means merely, but also as qualities of the life to be preserved and therefore as aspects of the end. It is one of the paradoxes of life that it employs means which modify the end and themselves become part of it (Jonas, 1982, p. 106).

Biological functions are not merely means to our pursuit of values, but become values in themselves. Loss of function is loss of value.

The Disvaluation Thesis

As we have seen, Engelhardt takes the concept of disease inherently to carry a negative valuation. Likewise, Whitbeck (1981a) takes the concept of disease to be inherently value-laden in the "capability sense." William Goosens (1980, p. 102) calls this claim of the essential value-ladenness of the concept of disease the "disvaluation thesis." While Goosens advocates a normative position on the concept of disease, he points out a problem with the disvaluation thesis. It appears that diseases can sometimes have benefits and so be valued. Examples such as cowpox providing immunity from the more serious smallpox, and sickle cell trait providing protection against malaria, are commonly given as evidence for this. The disvaluation thesis neither denies that diseases can have any benefit, nor does it claim that all diseases are on balance disvaluable. It merely contends that all diseases carry with them a disvalued component. His solution is to say that diseases are always some *threat* to well-being, with well-being to be construed broadly (Goosens, 1980, p. 107). This, as he recognizes, is still far too broad to be a complete account of disease, however. Many sorts of things, such as encounters with robbers in dark alleys, are threats to well-being but are not diseases. Goosens proposes that threat to well-being is a necessary condition for being a disease, but not a sufficient condition. He concludes that because the potential number of threats to well-being is indefinite, non-normative theories will never be able to provide a necessary condition for specifying health or disease (Goosens, 1980, pp. 113-114).

A problem for the disvaluation thesis is raised by the case where no disvalue seems to be connected with the disease. Goosens (1980, pp. 103-104) suggests the example of a disease that causes sterility in a person who desires not to have children. *Ex hypothesi*, the sterility is caused by a disease and so it is a medical condition. There is no question of balancing values and disvalues, as when the minor disadvantages of sickle cell trait confer immunity against the major disadvantages of malaria to yield a state of affairs having a net advantage. Sterility in the hypothetical case seems to carry no disvalue whatsoever. Therefore, under the disvaluation thesis we would have to conclude that the condition causing the sterility is not a disease. This contradicts the hypothesis of our thought experiment. Thus, if the disvaluation thesis holds, then sterility cannot be a disease.

Whitbeck's approach offers a solution. She holds that values enter into the concept of disease because diseases are the sorts of conditions that people would like to be able to control. This allows that a particular person with a particular disease might not disvalue it. All that is required is that people have an interest in being able to control the condition. Goosens thinks this approach fails to include normative concepts altogether. He says that Whitbeck's concepts of what persons "wish," "want," "expect," and "hope" are not normative concepts at all, but are merely empirical (and non-normative) descriptions of the states of persons (Goosens, 1980, p. 101). But this seems wrong. To wish to be able to control a condition is to make an evaluative assessment of the condition and one's relation to it. Saying that one *wishes* to allow a condition to progress or to take steps to eradicate it implies that one either values or disvalues it. To be sure, one can empirically describe the state of a person's wishing to control a condition. But one can empirically describe any value state. We can make empirical observations and state facts about any person's holding any value. This does not remove the evaluative component from such observations. Desires and wishes carry with them implicit judgments of value about what is worth pursuing and what is not.

In order to make this more explicit we recall Whitbeck's clause (1). Diseases are processes that "people wish *to be able to* prevent or terminate because they interfere with a person's ability to do things that people wish and expect to be able to do." To see this as an explicitly value-laden statement we need only to modify it as W. Miller Brown suggests. A disease is any type of psycho-physiological process that "people wish *to be able to* prevent or terminate because [it is undesirable in that] it interferes with the bearer's psycho-physiological capacity to do things that people commonly wish and expect to be able to do." (Brown, 1985, pp. 319-320).[6]

Sociocultural values may influence what is considered to be a disease. Whitbeck (1981a, p. 615) says that what qualifies as a disease is "relative to a societal context insofar as what people are understood as wanting to do is relative to societal context." Brown (1985, p. 321) objects to this claim, saying that if Whitbeck is correct, it will make no sense to say that what was once called a disease is now known not to be one. But to hold medical knowledge claims of the past to be folly in comparison with what we *now* know amounts to epistemological imperialism. Medicine in just the past century has seen many Kuhnian "paradigm shifts." The change in thinking about the etiology of Grave's disease from a primary hormonal disorder to an autoimmune disorder, and the redefinition of criteria for the diagnosis of rheumatoid arthritis are two examples. Alcoholism has been seen in different eras as a moral deficiency, a psychological dependency, or a physical disease that is in great part genetically determined. There is no reason to think that our concepts of particular diseases as well as what sorts of social deviancies are best explained and treated as diseases will not change in the future.

These shifts are often explained as being advances toward scientific truth. However, they might also be explained as advances in "value truth." The scientific evidence that appears to give support to a scientific realist view of the world, might in fact be an endorsement of the values that lie at the root of the facts. When everyone agrees on a fact, it is as much because everyone agrees on the values that are embedded in the fact as that they agree on the empirical aspects of the fact. Say, for example, that we all agree that a particular patient has pneumococcal pneumonia. We are agreeing about empirical observations on physical examination and laboratory testing. But we are also agreeing about (1) the foundational values that allow us to say that a particular morphological structure represents the pneumococcus, that the physical findings are sufficient to constitute pneumonia, etc., (2) the conceptual values that lead us to call pneumococcal pneumonia a disease, and (3) other values to be elucidated in the following chapters. This agreement is the heart of value-dependent realism.

IV. CONCLUSIONS

We began this chapter by considering the metaphysics of disease. The ontological conception of disease tends to see disease as a discrete entity, which is responsible for causing a patient's experience of illness. The classical ontological view saw disease as a sort of organism invading from outside, but later ontologists conceived of particular groups of cells within the body as constituting disease. The physiological conception of disease, on the other hand, tends to see disease as a disturbance in the normal balance of components within the body. This conception is more akin to present pathophysiological thinking, although it moves away from Sydenham's ideal of a nosology parallel to botanical classification.

Non-normativists think that disease can be defined without the use of value terms. To do this, they make use of the notion of natural functions, often relying on evolutionary theory to provide the basis to describe which functions are natural. However, functions are related to goals. Survival and reproductive advantage are the goals that non-normativists usually invoke, but there is good reason, as the normativists have shown, to deny that these goals are really value-free. The goals of an organism, especially one as complex as the human, cannot be specified without recourse to value.

We cannot remove values from the concept of disease. As I have argued in chapter 3, we cannot even remove values from the scientific facts that we observe and take as evidence of disease. There are, then, two levels of values operating in the concept of disease. *Foundational* values influence theory selection, choice of methods for empirical observation, statistical methods, choice of what is significant data and what is background noise, among other things. *Conceptual* values involve the judgments of societies and individuals about what constitutes a threat, what constitutes a harm, and what we would like to be able to control about our physiological nature. Where such values are universally or nearly universally held (and there are many such cases), we tend to see disease as a matter of pure fact. What this means is that in all disease there is a component of scientifically objective (although value-laden) facts, and also a component of conceptual values.

If this is true, and if diagnosis has something to do with explaining disease, then diagnosis will necessarily include the two types of values that are inherent in the concept of disease. In the next chapter, I will consider how diseases are classified, what counts as a disease, and what does not. More value judgments will be involved in choosing a classification scheme and in inserting particular conditions into that scheme.

CHAPTER 5

THE CLASSIFICATION OF DISEASES

I. INTRODUCTION

So far I have argued that the concept of disease is necessarily value-laden. Different sorts of values are involved at different levels in the construction of the concept of disease. Most basically, value judgments are made in the acceptance of any state of affairs as a scientific fact. These foundational values are the values that are epistemically fundamental for the scientific community, and they are usually so well accepted that they remain covert. The covert acceptance of these values is what can tend to make scientists think that scientific facts are simply discovered through observation that is value-neutral. As we go about the business of determining what is to count as a fact, the foundational values remain invisible, as the air we breathe remains invisible to us in the course of everyday life.

When we consider the concept of disease, we include a different set of values. These conceptual values, like the foundational values, are non-moral values. They are, however, more likely to be recognized as values by those in the scientific community for whom foundational values may be invisible. The debate about whether the concept of disease is essentially value-laden attests to this. Conceptual values have to do with specifying what sorts of bodily conditions and functions we favor or would rather do without, and with what states of affairs we would like to be able to control. In short, these specifications concern what we consider to be good and bad with respect to our bodily functioning. Once we speak of good and bad, we have entered the realm of values.

In this chapter I will consider the ways in which diseases are classified. We must take into account the empirical observation that clusters of scientific facts tend to occur together. We must examine the reasons why we see certain groups of facts as a cluster and other facts as unrelated. What I will show is that another set of values, *nosological* values, is involved in any system of disease classification. There are some relatively non-controversial epistemic values that come into play in disease classification. But in our classifications, we also make value judgments that reflect our purposes for making classifications and our interests in controlling what happens to our bodies. These nosological values come from the multiple realms of value we have already found at the foundational and conceptual levels. The foundational values and conceptual values are already contained in any entity we call a disease even before we classify it as a particular type of disease. Nosological values, however, constitute a separate *level of application* of value in the diagnostic process. That is, foundational, conceptual, and nosological values might all include values from several realms, such as scientific and aesthetic values. However, foundational, conceptual, and nosological values all play distinct roles in the classification of disease.

II. THE SCOPE OF NOSOLOGY

Although I have discussed several formal definitions of disease, I have not fully endorsed any one. That is, I have not attempted to give necessary and sufficient conditions for calling some state of affairs a disease. However, neither nosology (the classification of diseases) nor diagnosis requires such a precise definition. All adequate nosologies include entities that would not be considered diseases under one or another of the precise definitions. In fact, many conditions from which people suffer, and for which they seek help from physicians, are not considered diseases by the people who suffer from them. Diagnosis is just the attempt to translate all of these types of phenomena into the terms of some nosology.

The Domain of Medicine

The conditions that physicians have accepted as belonging to the domain of medicine include phenomena of many ambiguous and overlapping categories. Robert Veatch (1973b, p. 64) has demonstrated the tendency for a broad range of social deviancies to become incorporated into the medical model. A deviancy is put into a medical model if it is seen as (1) non-voluntary, (2) organic, (3) related to the expertise of the class of physicians, and (4) falling below some socially defined minimal standard of acceptable health.

Clauses (1) and (4) involve the sorts of conceptual values discussed in our consideration of the concept of disease. The non-voluntariness of certain diseases comes increasingly into question, as their etiologies become better understood. One might question whether or not a cigarette smoker who gets lung cancer and who knows that smoking is etiologically related to lung cancer has become afflicted non-voluntarily. However, one can always respond that the smoker did not voluntarily choose cancer, but only voluntarily chose to continue engaging in a known risk, hoping that the illness would not materialize in his or her particular case. In that case, the cancer itself is non-voluntary.

Conceptions of disease have social implications for what is considered to be within the proper domain of medicine. Horacio Fabrega (1972, pp. 584-598) suggests three independent frameworks for describing disease. First is the "biologistic" framework, in which specific diseases refer to biological processes. Second is the "behavioral" framework, in which diseases are recognized by "behavioral indicators" which are derived from social contexts. Third is the "phenomenological" framework, in which diseases depend on the experience of an altered conception of the self. Fabrega argues that these three frameworks are independent and logically self-contained, and that each can be used exclusively. However, the three frameworks share the view that disease occurs in individual organisms and is described as if individual diseases are discontinuous. Furthermore, all three frameworks take disease to be an undesirable condition. The social structures of the medical profession tend to be organized around these common assumptions (Fabrega, 1972, p. 599). There is no reason, however, to suppose that all patients and physicians are operating in the same framework. Even if there are common assumptions in all three frameworks, and even if all patients and physicians agree on these, there is no reason to assume that differences between the frameworks will not lead to important differences in perception about disease. As Fabrega recognizes, the social organization of medical practice around the common assumptions may mask the differences in underlying framework.

The Anatomico-Physiological Model

Veatch's clause (2) puts only organic deviancies into the medical model. As P. D. Toon suggests, it makes sense from a pragmatic point of view to classify diseases according to the way those who treat them are trained to deal with them. In our culture, physicians' training rests on the basic medical sciences. What the clinical and basic medical sciences share is what Toon calls the "anatomico-physiological model," a mechanistic model in which function is interpreted in terms of structure (Toon, 1981, p. 199).

Seeing disease in terms of this model accounts for the increasing medicalization of deviancy. Increasingly, researchers are finding specific genetic links to such behaviors as alcoholism, drug abuse and violent behavior, and even to such conditions as male pattern baldness and short or tall stature. As this happens, these conditions become more strongly incorporated into the medical model and enter into the sphere of expertise of medical professionals. Notice, however, that not all organic conditions are seen as diseases, but only those that deviate from an accepted norm, which is established according to conceptual values. However, there are some organic conditions that do deviate from accepted norms and are still not considered to be diseases. The growth of human hair is an organic phenomenon. One whose hair is dyed purple and fashioned into long spikes does not have a disease, at least not a disease of the scalp. A large and outlandish tattoo is not a disease, even though both the tattoo and the growth of hair involve organic phenomena. Purple hair and tattoos are not non-voluntary, and so they would not be diseases by Veatch's clause (1). However, the socially defined scope of expertise still plays an important role in our understanding of disease. It is, at present, not the physician's role to create tattoos, but it is the physician's role to remove them. Although the barber and the surgeon once were united in a single profession, the cutting and coloring of hair is no longer seen as being within the expertise of the medical profession. Thus, Veatch's clause (3) is an important factor in deciding which organic deviancies count as disease.

The Biomedical Problem

Illness, the subjective experience of being sick, becomes disease only when it acquires a medical explanation. There are many conditions that undisputedly fit the medical model according to Veatch's four criteria, but are generally considered not to be diseases. Injuries such as bone fractures and chronic conditions such as congenitally absent limbs are perhaps the best examples. Many people also seem to resist calling other common conditions, such as tension headaches and hernias, diseases. One suggestion of a term to encompass all the conditions that fit the medical model is "malady."

A person has a malady if and only if he or she has a condition, other than a rational belief or desire, such that he or she is suffering, or at increased risk of suffering, an evil (death, pain, disability, loss of freedom or opportunity, or loss of pleasure) in the absence of a distinct sustaining cause (Clouser, Culver, and Gert, 1981, p. 36).

Clouser, Culver and Gert (1981, p. 31) maintain that maladies are objective conditions, whether or not particular individuals or societies recognize them as conditions that need attention. Thus, they apparently exclude from their definition an element like Veatch's clause (3), which holds that the condition must be within the domain of medical professional expertise.

This definition of malady attempts to give an objective account of human illness while still affirming its value-ladenness. According to Clouser, Culver, and Gert (1981, p. 31), death, pain, disability, and loss of freedom, opportunity and pleasure are universally seen as evils; so, the concept of malady is essentially value-laden. Maladies, according to the authors, are intrinsically bad, although they can be instrumentally good. For example, when a soldier is sick at the time he is being considered for assignment to a dangerous mission, he has a bad condition that is instrumentally good.

The definition is not without its difficulties, however. First, it is not clear that the values asserted to be universal by the authors are indeed universal. For instance, death may not always be seen as an absolute evil, but merely as one disvalue to be balanced with others, as when one willingly dies as a martyr for some cause. This argument does not refute the claim that death is a universal disvalue. However, it does call into question the usefulness of asserting the objectivity of maladies, for being a malady in this sense is no warrant for treatment.

Second, the stipulation that the condition must occur in the absence of a distinct sustaining cause is meant to eliminate from the extension of the term "malady" such things as suffering caused by environments and circumstances such as extreme heat, which are distinct from the person. The test for determining whether a person is suffering from a malady is to see whether "changing the circumstances would rather quickly remove the suffering" (Clouser, Culver, and Gert, 1981, p. 33). So, according to the authors, the suffering caused by not having eaten in a few days is not a malady because a meal would quickly remedy the suffering. However, many infectious diseases are due to external circumstances, the infectious agent, and can be rapidly cured by a dose of antibiotics. Infectious diseases would appear to be exemplars of maladies. It will not do to say that the infectious disease is the body's response to the infectious agent and not the infectious agent itself, for hunger can be seen analogously as the body's response to lack of food. Any disease may or may not be considered to be distinct from the person depending on one's ontological view of disease.

The discussion of the concept of malady points out that in classifying the illnesses that patients bring to them, physicians are dealing with more than strictly defined diseases. Any adequate nosology will include a broad range of conditions; it will include all such matters that a society recognizes as part of the medical model and therefore brings to medical professionals for treatment. Because of the problems with the concept of malady, Engelhardt (1996. pp. 205-206) has proposed the term "clinical problem" to make explicit that a broader category of entities fall within the purview of the medical profession. This term underscores the value-ladenness of the concept in asserting that what is brought to a physician is seen as a problem. A difficulty with it, however, is that it apparently suggests a connection with the clinic (broadly construed as clinic, hospital, doctor's office, etc.). But this connection may not hold for many of the things we rightly consider diseases, maladies, or clinical problems. Such things as the common cold, strained muscles, and minor cuts and scrapes rarely come to the attention of professionals in the clinic. They do, however, retain the character of non-voluntary, organic, disvalued conditions that are related to the expertise of the medical profession.

This is not a very powerful criticism of Engelhardt's term. It might be argued that people with headaches and muscle strains often do come to the clinic, and not inappropriately. Nonetheless, I would prefer to de-emphasize any necessary connection with the clinic, but retain the elements of being non-voluntary, organic, disvalued, and related to the expertise of the medical profession. I suggest the term "biomedical problem." The term "disease," however, is so prevalent in the literature that I will follow these authors in using "disease" to mean "biomedical problem." If a narrower sense of "disease" is called for, I will specify that. Note also that mental illnesses may properly be categorized as biomedical problems when they fulfill Veatch's criteria. I have, however, chosen not to discuss these in this book.

The Nature of Nosologic Categories

There are various ways to construct a nosology. The way that is chosen depends on what is considered to be most important. Constellations of symptoms and signs seen as falling into natural kinds, constructions according to some societal values, and divisions according to particular types of medical expertise are all possibilities. The classification that is ultimately accepted will depend upon facts about the conditions classified, but it will also importantly depend on values that determine which of all the possible classifications is chosen.

The choice of a nosology will have important ramifications, not only medical but also social and economic. Nosology is often forced into service as a *lingua franca* for different worldviews. A result is the medicalization of various sorts of conditions.

> [D]isease classifications serve to rationalize, medicate, and legitimate relationships between individuals and institutions in a bureaucratic society. This is nicely exemplified in third-party payment schemes, where the inchoate and possibly incommensurable experiences of individuals are transformed into the neatly ordered category of a diagnostic table—and thus suitable for bureaucratic use. In this sense a nosological table is a kind of Rosetta Stone providing a basis for translation between two very different yet stucturally interdependent realms (Rosenberg, 1992, p. xxi).

In the remainder of this chapter, I will consider some of the classic nosologies that have been in use, and examine the ways in which different types of biomedical problems are constructed. I will show how these nosologies reflect particular facts and nosological values.

III. THE CONSTRUCTION OF CLASSIFICATIONS

Even if some diseases are discrete natural kinds, all classifications of biomedical problems are human constructions. In the construction project, another layer of values is added to the foundational values already involved in deciding that any given observation about a sick person is a scientific fact and the conceptual values that go into understanding the nature of a biomedical problem or disease.

Edmond Murphy (1976, pp. 101-108) has suggested that there are six desirable characteristics of a system of classification. Note that he calls these characteristics desirable. This tells us that we have entered into the realm of values. I take Murphy's six characteristics as fundamental nosological values. We will encounter others as we go along. Murphy's characteristics are as follows:

A. *Naturalness*. A classification should correspond to the nature of the thing being classified. First of all, notice that the word "should" is used here; this itself indicates a normative judgment. This characteristic suggests that "natural" is being used in the sense of the naïve realist, that classifications should correspond to the way things are. But Murphy is subtler than this. He is very cautious here, recognizing that an individual person, *qua* individual, cannot be classified without distortion, and that medical practice is certainly more than putting sick people into categories. Nonetheless, sufficient similarities recur so regularly in certain sorts of biomedical problems to suggest that there are natural categories, value-laden though they may be, to which our constructed classifications ought to correspond.

B. *Exhaustiveness*. A classification ought to put every member of the sample set for the particular issue into some class. A complete classification for medicine should include all the illnesses and conditions for which people seek medical help. This suggests that diagnosis will be a constructive process, adding new classifications when necessary, rather than just a process of sorting individual cases into previously determined categories. Particular classifications, however, need only concern particular problems. For example, a classification of gynecological diseases need not include diseases of the prostate.

C. *Disjointness*. Classes should be disjoint. That is, no particular case should fall into more than one class. We will see that this ideal is problematic. Many diseases have a multifactorial etiology and significant overlap, and may be better thought of in terms of continuous dimensions rather than as discrete categories (Toon, 1981, p. 200), or as two or more separate categories.

D. *Usefulness*. Murphy rightly contends that usefulness is the most important property to include in a system of classification. Whether an element is useful for classification will depend on the reasons one has for making the classification in the first place. Usefulness for the medical researcher, the clinician, and the patient may be very different. Murphy

(1976, p. 104) suggests five possibilities of classification according to usefulness:

1. the basic mechanisms of the disease (e.g., tuberculosis, the hemoglobinopathies);
2. descriptive features (e.g., syndromes, many skin diseases, most diseases of the heart);
3. effective means of treatment or management (e.g., arthritis, asthma);
4. prognosis (e.g., leukemia, multiple sclerosis);
5. the convenience of other disciplines, such as hospital administration, the law, education (e.g., mental deficiency, blindness).

He contends that rationality for treatment will improve when classifications lower on the list yield to those higher on the list. It is here that value judgments in classification become most important. There are, of course, many other ways to classify disease. I will consider this aspect in more detail later.

E. *Simplicity*. In general, the simpler a classification is, the easier it will be to use. This is a pragmatic value akin to the value of simplicity in choice of a scientific theory.

F. *Constructability*. A system of classification is constructable if its structure allows the satisfaction of both exhaustiveness and disjointness. This is an ideal that can sometimes be met by stipulation that a particular ambiguous case belongs to one category and not another. However, as we will see later, classifications complex enough to account for current medical practice will not meet this ideal of constructability.

As Murphy recognizes, conflict among these various ideals in any classification scheme is inevitable. This conflict is especially evident between naturalness and disjointness. Murphy (1976, p. 108) sees this conflict as the reason for most misclassifications. This notion of "misclassification" seems to put him into the scientific realist camp, as it implicitly asserts that basic disease mechanisms are the basis for the one correct classification. However, Murphy sees all of our knowledge as tentative and further recognizes the difficulties we have in separating relevant facts from background noise in classification. He thus also appears to lean toward epistemological skepticism.

Murphy (1979, pp. 59-85) also suggests two alternatives to classification: clustering and measurement according to some established scale.

In classification, a domain is ideally broken into categories that possess necessary and sufficient conditions for deciding whether a particular element belongs to the category. An element is classified if it possesses the necessary and sufficient conditions for inclusion. There may be elements that do not possess all the conditions, however, but whose conditions tend to cluster together, bearing only a Wittgensteinian family resemblance to one another. Clustering has been criticized as nothing more than a type of classification in which no necessary and sufficient conditions can be stipulated (Beckner, 1979, pp. 87-92). This idea of clustering, however, seems better to present the case for categorizing almost all diseases. Carl Hempel (1965, pp. 151-152) remarks:

> Classification, strictly speaking, is a yes-or-no affair: A class is determined by some concept representing its defining characteristics, and a given object falls either into this class or outside, depending on whether it has or lacks the defining characteristics.
>
> In scientific research, however, the objects under study are often found to resist a tidy pigeonholing of this kind. More precisely: those characteristics of the subject matter which, in the given context of investigation, suggest themselves as a fruitful basis of classification often cannot well be treated as properties which a given object *either* has *or* lacks; rather, they have the character of traits which are capable of gradations, and which a given object may therefore exhibit *more or less* markedly. As a result, some of the objects under study will present the investigator with borderline cases, which do not fit unequivocally into one or another of several neatly bounded compartments, but which exhibit to some degree the characteristics of *different* classes.

In the absence of necessary and sufficient conditions to stipulate that a given state of affairs belongs to some classificatory set, it will be necessary to make value judgments about how much of a family resemblance is necessary to judge that one is dealing with a cluster that constitutes a disease category. Additional value judgments about the usefulness of one categorization over another are likely to enter into nosology. Such value judgments about usefulness are likely to reflect conceptual values about what sorts of conditions are disvalued and are therefore considered to be diseases.

I will later turn to a detailed analysis of the problems and conflicts of ideals in the classification of particular kinds of biomedical problems.

First, however, I want to consider a few historically important attempts at disease classification and examine how fact and value interact in them.

IV. EXAMPLES OF NOSOLOGIES

Nosologies of Historical Interest
My intention in this section is neither to trace comprehensively the history of nosology, nor to do any primary exegesis of historical medical texts. My interest lies merely in illustrating some of the possible ways to classify biomedical problems. The history of nosology shows that many possibilities for classifying diseases have actually been advocated during the past four hundred years.[1]

Classification by Clinical Observations
In the seventeenth century, Thomas Sydenham (1979) advocated an ontological conception of disease, believing that all diseases were of a certain definite species. He thought that philosophical speculation should be set aside, that clinical observation would be sufficient to elucidate the natural history of disease, and that this would provide the needed information for a complete classification of diseases (Faber, 1923, p. 7).

In the eighteenth century, under Sydenham's influence, François Boissier de Sauvages (1768) constructed a nosology of twenty-four hundred diseases, with ten classes and forty-two orders, the orders further divided into genera and species. The classes of disease were based strictly on clinical criteria such as fevers, spasms, pains, and wastings, reflecting the virtual non-existence of any sort of technologic means for diagnosis.

Such nosologies sought explicitly to be exhaustive in scope. Given the assumption that diseases were ontological kinds, they implicitly sought to affirm the ideal of naturalness. However, in rejecting philosophical speculation in favor of empirical observation they make a hidden value judgment about the superiority of the empirical powers of the methods available to them over rationalistic methods for establishing an exhaustive nosology. Furthermore, this method of typology was chosen because of the belief that the resulting classification alone would be useful in providing a basis for rational treatment. Facts about empirical observations are grouped in these classifications according to nosological value judgments about epistemological methods and usefulness for a particular goal, treatment.

Anatomico-Clinical Classification

As various methods were found to look into the body, new conceptions of disease developed. Post-mortem examinations of the body revealed that particular organs exhibited lesions, which could be correlated with the clinical manifestations observed during life. In addition, means were found to look into the body of a living sick person. Anatomic lesions came to be considered to be the essence of disease. René Laennec (1781-1826), inventor of the stethoscope, is regarded as one of the founders of the anatomico-clinical school. Laennec recognized particular heart sounds as correlates of particular anatomic lesions in the heart. Furthermore, he recognized that such lesions could be present in asymptomatic patients, an early recognition of Feinstein's "lanthanic" patients (Feinstein, 1967, p. 145). Hence, Laennec (1982, MS. 2186 (III)) showed that nosologies based only on a clinical description of symptoms would omit this group of patients and would not be exhaustive.

Laennec also realized that not all disease could be reduced to a lesion in a solid organ. He therefore developed a conception of the human organism as consisting of three parts: solids, liquids, and the *principe vital*, an animating force. All diseases were classified as lesions in one of these parts. For example, hernias and fractures were lesions of solids, fevers of various sorts and poisonings were lesions of liquids, and epilepsy, coma and dementia were lesions of the *principe vital* (Duffin, 1986, pp. 195-219).

Sydenham and Laennec approach the task of classifying diseases from different worldviews. Both might accept Murphy's six ideals for classification. However, their different worldviews could still lead them to construct radically different nosologies. Sydenham's worldview rejects any means of construction that does not depend on empirical observation. Laennec begins by embracing this view and even extending it to include means of observation not available to Sydenham. However, he quickly realized that Sydenham's method fails the test of exhaustiveness when so extended. Some diseases manifest no symptoms, and some diseases have no discrete lesions in either solid organs or liquids. Laennec was forced to adopt the *principe vital*, just the sort of philosophical construction that Sydenham eschewed.

The anatomico-clinical ideal is taken to the extreme by the nineteenth century German pathologists. Rokitansky's nosology of 1876 holds that pathological anatomy should be the basis for both the knowledge and practice of physicians (Faber, 1923, p. 57). This view is dominant to this day in the clinico-pathologic conference, where the pathologist takes the podium after a problematic case has been thoroughly discussed and provides the true diagnosis, which has often eluded the clinicians.

Classification by Etiology

The microbiologists of the late nineteenth century established that microorganisms are the etiologic agents in a vast number of diseases. Infectious diseases are perhaps the best exemplar of disease construed in a narrow sense. In classifying disease by etiologic agent, one emphasizes simplicity and naturalness. Such a classification can be very useful in deciding on a rational method of treatment. In some ways, the work of the microbiologists vindicated Sydenham's ontological conception of disease. The primary difficulty with classification by etiology, however, is that it leaves unclassified many biomedical problems whose etiology remains unknown. Furthermore, it may fail to discriminate between two conditions that have a common etiologic agent but different manifestations. For instance, group A streptococci can cause a variety of biomedical problems, from pharyngitis to impetigo to necrotizing fasciitis to puerperal fever. Classifying all these conditions by etiologic agent gives some rational basis for treatment but does not account for the variety of clinical manifestations, anatomic lesions, or the necessary variations in indicated treatment for these conditions.

Functional Classification

As time went on, technological means to carry out clinical research were developed and became more complex. Even such simple devices as Kussmaul's stomach pump made it possible to describe physiologic functions in ways that were not before possible (Faber, 1923, p. 112). Classifications according to function favor physiological conceptions of disease over ontological ones. Faber (1923, p. 164) notes:

> Functional diagnosis was introduced with the watchword "*Wir wollen heilen und nicht klassifizieren,*" but it has shared the fate of all other clinical methods; they all lead ultimately to new classification, to the dissociation of non-homogenous processes, and to the consolidation of homogeneous ones.

Functional classifications include conditions that lack a clear or simply described etiologic agent as well as conditions that are not associated with any particular gross anatomic lesions. They admit biomedical problems that might not be admitted under a more formal and strict definition of disease.

My intent in setting forth these various historical nosologies has not been to point out their deficiencies. That would simply be to claim that the lack of technological ability of past generations did not allow them to see things in the way we now see them. Rather, the point is that different worldviews can lead to different ways of classifying any biomedical problem. One's choice of classification schemes depends largely upon which philosophical presuppositions one is not willing to relinquish, as well as the goals one has for making the classification.

Current Nosologies

I turn now to classifications of disease in current use. What is characteristic of these nosologies is their inclusion of elements of all of the above methods. This way of proceeding reflects a worldview that places high value on exhaustiveness at the cost of simplicity.

It is tempting to view competing schemes of classification merely as alternative ways to categorize a finite set of diseases, like alternatively sorting a set of objects first by color and then by size. Recall, however, our discussion in chapter 2 of Putnam's example (1987, pp. 18-20) of how the number of objects in a defined universe can be understood differently in different worldviews. Different classifications may reflect similar differences of worldviews. Take the simple example of an individual with two well-defined diseases. The effects of these two diseases on the individual may not be simply additive. The effects of one disorder may serve to mitigate those of another, as in the case of sickle cell trait and malaria. Alternatively, the simultaneous occurrence of two disorders such as influenza and emphysema may result in a process that causes higher morbidity than a simple sum of the morbidity of the two conditions. In such cases, it is not obvious that we should say that the individual simply fits simultaneously into two disease categories rather than into some third category that is yet undefined.

Systematized Nomenclature of Pathology (SNOP)
The Systematized Nomenclature of Pathology (SNOP) is an attempt, as its name suggests, to standardize the nomenclature used by pathologists in diagnosis.[2] It is intended to be an aid to pathologists for the cataloguing of specimens. While its domain consists of specimens examined by the

pathologist in the laboratory, such specimens represent virtually the entire gamut of human disease. Thus, while SNOP makes no claim to be clinically oriented, it attempts to be exhaustive insofar as the biomedical problems catalogued are represented by some sort of aberration in a body tissue or fluid. That is, SNOP is exhaustive given the anatomico-clinical model, which reflects the materialist worldview of all the natural sciences. There is no place in SNOP for lesions of Laennec's *principe vital*.

SNOP takes diseases to be describable with respect to four fields: the part of the body affected (Topography), the structural changes produced (Morphology), the etiologic agent (Etiology), and the functional manifestations (Function). Within each field, terms are assigned a four-digit number, with the first digit representing the section of the field and the other numbers indicating finer subdivisions. SNOP takes morphology to be of primary importance in describing a disease from a pathological standpoint. Hence, if one is interested in reviewing all cases in which inflammation plays a role, one can retrieve all cases assigned the number M-4000, the code for "Inflammation, not otherwise specified." When a topographic site such as colon (T-67) is added, the resulting code, T67-M4000, indicates "colitis." The etiologic field includes various pathogenic organisms as well as chemicals and drugs that cause disease and injury. The function field covers a variety of physiological and chemical states as well as a number of normal states such as pregnancy, and even some complex disease entities. Its purpose is to allow for a complete coding and it allows one to distinguish morphological changes with and without functional evidence of disease, as well as functional evidence of disease without morphological changes. A case in which ulceration (M4003) of the ileum (T65) was observed due to typhoid fever (F9497) and in which *Salmonella typhi* was recovered (E1361) would be coded as T65-M4003-E1361-F9497. If the bacterium were not recovered, the E field would be omitted; if the functional evidence, i.e., the clinical signs, of the disease were not present, the F field would be omitted. Coding examined specimens makes possible future recovery of records, tissue slides, photographs, and so forth. It also aims at providing a standard nomenclature, which may help reduce ambiguity resulting from use of synonymous medical terms for a single condition.

The use of a nomenclature such as SNOP relies on virtually all the elements of the historical classifications already discussed, including anatomic, etiologic and functional classifications. By omitting particular fields, one can indicate that that particular field is unimportant in the classification of a particular case. The fact that morphology is considered to be the heart of SNOP, however, does not reflect the relative importance of morphology for describing disease as much as it reflects the centrality of morphology in the work of the pathologist.

Systematized Nomenclature of Medicine (SNOMED)

The Systematized Nomenclature of Medicine (SNOMED) is an extension of the "SNOP concept"; it is said to be "derived directly from the philosophical study of the nature of man and gives us the basic axes or categories of the nomenclature" (College of American Pathologists, 1979, p. xvi). SNOMED contains the four fields of SNOP and three additional fields. The disease field was added as a way to include discharge diagnosis in medical records for statistical reporting purposes. The disease field, however, cannot be used alone to report specific components of complex diseases. It is reported to be "in reality a classification of diseases ... created as a logical extension of a basic four-axis, hierarchically structured and philosophically sound nomenclature of medicine" (College of American Pathologists, 1979, p. 375). The compilation of the list of diseases was "inspired by standard textbooks of clinical medicine, recent subject review articles, and numerous personal contributions by our consultants" (College of American Pathologists, 1979, p. 376). A distinction is thus made between a nomenclature, which merely purports to name observations, and a classification, which imposes order on a set of complex phenomena. The disease field is the classification component of SNOMED. It should also be noted that SNOMED's authoritative source for classification is vademecum science, the distillation of medical research into textbooks and review articles. Appeal to this particular authority reflects the epistemic and professional values of the authors.

The sixth field in SNOMED is the procedure field. This field includes a list of administrative, diagnostic, therapeutic and preventive procedures. The major use of the procedure field would be more related to administrative purposes than to the understanding and classification of disease.

The seventh field is the occupation field. It uses an international standard of classification of occupations from the International Labor Organization. The rationale for including occupation as a separate field is to provide a formal means of reporting a patient's work for relevant specialties such as industrial medicine.

The inclusion in SNOMED of procedure and occupation fields reflects the complex administrative needs of medical record keeping more than it enhances understanding of disease and disease classification. The disease field, on the other hand, takes SNOMED a step beyond SNOP. The "sound philosophical principles" upon which this classification is built are nowhere made explicit. The strongest appeal is to the classifications found in standard textbooks of medicine. Thus, the classification of SNOMED apparently depends on the preferences of textbook authors whose reasons for organizing material in a particular way may depend as much on the values of effective pedagogy as on effective therapy.

International Classification of Diseases (ICD-9)
The ninth and most recent revision of the International Classification of Diseases of the World Health Organization (ICD-9) explicitly recognizes the many possible choices for classifying "morbid entities." When there are multiple axes of classification within a system, the particular axis chosen will be determined by the interests of the one using the system. While recognizing that no single classification will fit all specialized needs, ICD-9 attempts to provide "a common basis of classification for general statistical use; that is, storage, retrieval and tabulation of data" (World Health Organization, 1977, p. vii). Various adaptations of ICD for different interests (oncology, for example) have also been released. Updates to these classifications are published regularly.

ICD-9 appeals to the ideals of exhaustiveness and disjointness in its assertion that "every disease or morbid condition ... must have a definite and appropriate place as an inclusion in one of the categories of the statistical classification" and that "miscellaneous categories should be kept to a minimum" (World Health Organization, 1977, p. vii).

In ICD-9, diseases are given unique three digit numbers, with additional digits further specifying the disease. For instance, malignant neoplasm of the stomach is 151, with 151.0 designating a neoplasm of the cardia, 151.1 a neoplasm of the pylorus, and so forth. The diseases are divided into seventeen broad categories. The categories, however, are heterogeneous and do not represent any consistent way of categorizing disease. Eight refer to anatomic sites: blood and blood-forming organs, nervous system and sense organs, circulatory system, respiratory system, digestive system, genitourinary system, skin and subcutaneous tissue, and musculo-skeletal and connective tissue. Two refer to etiology: infectious and parasitic diseases, and injury and poisoning. Four refer to functioning: neoplasms;[3] endocrine, nutritional and metabolic diseases and immune disorders; mental disorders; and complications of pregnancy, childbirth, and puerperum. Two categories refer primarily to time of onset: congenital disorders, and perinatal disorders. One is a miscellaneous category for symptoms and signs that are not labeled as belonging to any well-defined disease. In addition to these seventeen, there are supplementary classifications: an "E" field for external causes of injury or poisoning, a "V" field for factors influencing health status and contact with health services, and an "M" field for morphology of neoplasms. There are also cross-references for some diseases. However, in such cross-references, one entry (marked by a dagger) is considered to be primary, and the other (marked by an asterisk) is considered to be secondary. For example, the primary classification of acute poliomyelitis is found under infectious and parasitic diseases, entry 045; a secondary cross-reference to poliomyelitis is given as 323.2 in the nervous system and sense organs category.

ICD-9 reflects an explicit commitment to exhaustiveness in that it seeks to classify the entire spectrum of disease; it also reflects a commitment to seeing the entire spectrum of disease in particular patterns. However, it makes no consistent use of patterns, as did the historical classifications based on particular ideologies of disease. Instead, ICD-9 includes bits of all of these ideologies.

V. CONSTRUCTING DISEASES FROM FACTS AND VALUES

Realism vs. Social Constructivism, Again

The preceding classifications of disease seem to presume that there are naturally occurring ontological kinds of disease, and that the fundamental task of the nosologist is to exhaustively sort them into categories that are consistent with these kinds. The further presumption is that just such a classification will be useful for diagnosis, prognosis and treatment. Problems of miscellaneous symptoms that defy categorization simply await further elucidation from research that will allow their proper categorization. Diseases that can be cross-referenced merely reflect the fact that some objects can be sorted according to more than one useful criterion. What I have suggested, however, is that the notion of an object is not a simple one, and this becomes especially evident when the objects are biomedical problems. Biomedical problems are constructed out of sets of signs and symptoms. These "building blocks" are facts (which are themselves value-laden), values, and sometimes even other biomedical problems. The construction project leads to some difficulties, and requires further value judgments to decide how to resolve them.

A radical expression of the social constructivist view of disease classification is that of Ivan Illich (1976, pp. 166-169, passim), who argues:

> Medical epistemology ... will have to clarify the logical status and the social nature of diagnosis and therapy, primarily in physical–as opposed to mental–sickness. All disease is a socially created reality ... .
> ... Substantive disease can thus be interpreted as the materialization of a politically convenient myth, which takes on substance within the individual's body when this body is in rebellion against the demands that industrial society makes upon it.
> In every society the classification of disease–the nosology–mirrors social organization. The sickness that society produces is baptized by the doctor with names that bureaucrats cherish.

Illich makes disease into nothing more than a social construction to serve a political agenda. Disease classifications, however, are not wholly arbitrary. Neither are they wholly dependent on the interests of bureaucrats. As Lawrie Reznek argues, the class of all diseases does not represent a natural kind. That is, there is no explanation of the clustering of all the phenomena we call "disease" according to the laws of nature. However, there may be naturally occurring clusters of phenomena that we regularly observe and call one disease or another. Even though the class "disease" does not represent a natural kind, at least some particular diseases may be natural kinds, and some of our disease language may refer to such diseases with what Reznek (1987, p. 49) calls "natural kind semantics."

Diseases and Natural Kinds

Reznek's theory of natural kind semantics is what he calls a combined theory. Objects belong to the same kind if they satisfy some minimal description of the kind as a whole. This alone, however, is insufficient to fix the extension of the natural kind. For example, suppose tigers were taken to be a natural kind by virtue of some descriptive semantics, say, that tigers are large striped cats. Then, if a particular tiger failed to acquire stripes because of a genetic disease, it would not be a tiger. Thus, neither the nominal essence theory of semantics of Frege (1949) and Russell (1985) nor the family resemblance theory of Wittgenstein (1973, §67) will suffice to fix the extension of natural kinds. This situation is just what we have with particular diseases, for typically very few particular cases of a disease fit a description of the "classic case."

Reznek's combined theory adds a causal component akin to the semantic theories developed by Saul Kripke (1980) and Hilary Putnam (1975, pp. 215-271). In these accounts, something is water if and only if it stands in the right sort of causal relationship (in this case, being H_2O) with samples of the stuff we refer to as "water."

In the Putnam-Kripke view, all members of a natural kind possess some constitution that makes the natural kind what it is, and that constitution is empirically discoverable through the methods of natural science. This metaphysical view has been criticized as merely reflecting a "culturally and philosophically specific *Weltanschauung*" (Cassam, 1986, p. 98), gratuitously assuming the existence of essences (Mellor, 1977, p. 309), and of making the extension of many of the natural kind terms we presently use in ordinary language depend on future scientific discoveries (Donnellan, 1983, pp. 102-104). Robert D'Amico (1995, pp. 554-555) argues that Reznek's picture of natural kinds rests solely on this type of essentialism, in which the notion of a kind refers to the "lawlike relations that hold for the essential or microphysical structure of things." Reznek, however, responds that our classifications of disease reflect not so much their causes, but their effects. Pathological states as a whole are grouped together not because they are a natural kind, but because we have an interest in avoiding them (Reznek, 1995, pp. 581-582). This does not say that no diseases are natural kinds, but only that the class of diseases as a whole is not a natural kind.

I think Reznek is right about this. It will become obvious as we consider some examples of particular diseases that they are not adequately described as only natural kinds in the Putnam-Kripke sense. As John Dupré has argued, even the complexities of taxonomy of biological species often will not allow such precise definitions of biological essences as are required by the Putnam-Kripke view. Dupré's position on taxonomic classifications, which he calls "promiscuous realism," illustrates this.

The realism derives from the fact that there are many sameness relations that serve to distinguish classes of organisms in ways that are relevant to various concerns; the promiscuity derives from the fact that none of these relations is privileged. The class of trees, for example, is just as real as the class of angiosperms; it is just that we have different reasons for distinguishing them (Dupré, 1981, p. 82).

Value-dependent realism is similar in its holding that there are many ways to distinguish biological processes, but it tries to avoid promiscuity in its assertion that an appeal to value realism can justify the claim that some classifications are objectively better than others.

The importance of including both descriptive and causal components in natural kind semantics becomes evident when it is discovered that some class of objects formerly thought to be a natural kind does not possess the proper sort of causal relationships. When that happens, one possible response is to conclude that the natural kind no longer exists. But, as Reznek points out, more often what we do is to change the meanings of the terms involved. We thus need a descriptive component (Reznek, 1987, p. 59). It is important to recognize that deciding when to make such changes will be a value judgment, and will probably reflect values on several different levels.

Were we to take disease as a natural kind of the Putnam-Kripke sort, we would be forced to omit much of what is important to people in their experience and understanding of illness. To take tuberculosis to be essentially described as infection with *Mycobacterium tuberculosis* and a particular immunologic response is to miss many of the features of the disease as experienced, the suffering of Mimi in *La Bohème* and Little Eva in *Uncle Tom's Cabin*.[4] The tuberculosis described in the great literature of the nineteenth century constitutes a disease kind that does not wholly depend on its description in scientific terms.

The difference between illness and disease is often seen simply as the difference between a patient's experience and a physician's scientific description of the same phenomenon. As S. Kay Toombs (1992, p. 24) shows, illness as subjective experience cannot wholly be shared with another, not only because of the failure of "interchangeability of standpoints," but also because of "incongruence in the appresentational schema of physician and patient." This gap between illness and disease is not an unbridgeable chasm. A patient, in fact, may experience illness as a disease; this is especially the case for patients who are scientifically sophisticated. A person with a family history of heart disease may experience chest pain as "having a heart attack" whereas the person without the same family history may simply experience pain (Toombs, 1992, p. 36). Diseases, understood as natural kinds of the causal sort, may partially account for the phenomena that are the object of diagnosis, but they do not wholly account for them.

Types of Biomedical Problems

I will now consider several different types of the biomedical problems and examine how they have been constructed and how the construction process incorporates nosological values. Henrik Wulff (1976, p. 67) has divided disease types into the four categories: symptom diagnoses, syndromes, anatomically defined diseases, and causally defined diseases. This will be a useful starting point, but we will see that these categories admit some ambiguities. There are biomedical problems that can fit into several of these categories. Understanding one as primary involves a nosological value judgment, which reflects the various purposes we have for categorizing biomedical problems.

Symptom Diagnoses
Wulff (1976, p. 68) suggests that symptom diagnoses are the "most primitive form of disease entity." This type of biomedical problem consists of a single symptom or clinical sign or finding. Wulff says that such diagnoses are often preliminary and "subordinate to other types of disease entities," and that with the addition of more information they can be turned into types that are more specific. He gives the example of chronic diarrhea, which could turn out to be cancer of the colon (anatomic diagnosis), ulcerative colitis (syndrome), or lactase deficiency (causally defined disease).

Terra Ziporyn (1992, pp. 99-101), while admitting that "disease" is an abstraction and has no precise definition, cautions against mistaking symptoms for diseases. For her, a symptom diagnosis can cause undue worry in a patient, or dismissal of a patient's complaints by a physician, as in her example of a post-cataract patient who is told she has "aphakia."[5]

Mere descriptions of symptoms are often the stuff of "admitting diagnoses"; however, they are generally not looked upon as definitive diseases, but merely problems to be further elucidated. However, sometimes symptom diagnoses are seen as definitive. Dermatology offers

many examples. The old joke defines a dermatologist as a doctor who tells you in Latin what you just told him in English. There is a kernel of truth in this. A patient may present with a greasy, scaly red rash, and go away with a diagnosis of seborrheic dermatitis. A patient with an itchy rash might enjoy some psychological comfort (or distress) in having a diagnosis of eczema rather than just a rash. However, this type of symptom diagnosis *per se* tells virtually nothing about such things as cause and prognosis, which usually are the point of making diagnoses in the first place.

Nonetheless, symptom diagnoses still may be useful. Suppose a patient presents with a rash of a certain sort that can be empirically examined and classified according to agreed-upon standards. That the patient has a rash is a fact, although many foundational values, epistemic and aesthetic, for example, have come into play in calling a particular skin phenomenon a rash. When such values are well accepted, we accept the fact about the rash. However, there is no symptom diagnosis without the addition of values to these facts. First, there are conceptual values concerning whether the facts about this rash point to the non-voluntary, organic, disvalued, and related-to-the-medical-profession criteria that would make it a disease. Second, there are nosological values that make the rash a particular kind of disease. Usefulness, which is a value judgment, is often what turns the facts about a disease into a particular diagnosis. Certain rashes are called eczema because that is a useful way to characterize them for purposes of treatment with a topical glucocorticoid. Preliminary diagnoses such as chronic diarrhea may be useful in steering further investigation. Such a diagnosis, of course, may also steer further investigation in a wrong direction, that is, one that is not useful. Nonetheless, all such judgments incorporate nosological values.

Anatomically Defined Diseases

Wulff's example of an anatomically defined disease is cancer of the colon. However, while the anatomic site of the cancer is part of the definition of colon cancer, it is not what we take to be most important. The most important aspect of colon cancer lies in its functional aspects: its tendency to metastasize. Inflammatory masses in the colon can sometimes appear very similar to cancers on gross anatomic inspection. It is very important to distinguish between such masses for purposes of treatment and prognosis. Thus, it is not primarily anatomic site but function that is of chief importance in colon cancer.

This is not to say that a classification of diseases by anatomic site is not useful. Seeing colon cancer as an anatomically defined disease can actually be quite important in differential diagnosis, the process of drawing up a list of possible conditions with which a set of signs and symptoms is consistent, and coming to a conclusion about which of these conditions the patient actually has. At the stage of making a diagnosis, finding a lesion in a particular anatomic site can be the most important classifying feature.

A better example of an anatomically defined biomedical problem (although not a disease, strictly understood) is an injury. Bone fractures are primarily defined by their anatomic site and by comparison with what constitutes the normal anatomy of the bone, that is, its physical continuity. Likewise, a ruptured spleen is primarily defined by its anatomic site. The importance of surgical treatment for this condition stems largely from knowledge of its anatomical character of being highly vascular and susceptible to massive hemorrhage when ruptured.

Causally Defined Diseases

Understanding and classifying diseases according to their causation is in some ways the most complex process in classification. Part of the reason for this complexity lies in the philosophical difficulties involved in understanding the notion of causation.[6] Causal classification, however, is often the most promising way of classifying disease if one is primarily interested in providing a rational basis for treatment. Causally defining a disease requires knowledge of its etiology as well as the normal and pathological functions of the body or part of the body affected.

The infectious diseases are probably the clearest examples of causally defined diseases. Pneumococcal pneumonia, for example, is defined by presence of the pneumococcus in the lung.[7] The presence of the etiologic agent, however, is a necessary but not sufficient condition. The pneumococcus can be present in the lung without causing pneumonia. The presence of pneumonia depends on a complex of factors including the etiologic agent, but also including characteristics of the host and the environment. The presence of the etiologic agent is only one necessary part of a complex of parts, which are jointly sufficient to cause the disease. The etiologic agent, then, is not the total cause of the disease; it functions more like one of John Mackie's "inus" conditions.[8] The complex of parts are sufficient to cause the disease, but are not necessary insofar as the disease might also arise given a different set of factors.

However, given the sufficient set of factors, the etiologic agent is a necessary component of the set.

Causally defined diseases, then, do not rest on a commitment to an ontological view of disease. In causally defining a disease, one is not necessarily saying that the disease is to be equated with the etiology. One is merely holding that etiology is the most important factor for classifying the disease. What makes the etiology of infectious diseases important from a classificatory standpoint is that the etiology explains the disease in ways symptom diagnoses or anatomically defined diseases cannot. This is useful in a couple of ways. First, by isolating a necessary (but insufficient) cause of a particular disease, one can tailor treatments that aim at eliminating the necessary cause, that is, the etiologic agent. Second, knowing that a particular disease is caused by some etiologic agent may allow the prevention of the disease by eliminating the opportunity for human-agent contact.

Another example of a causally defined disease depends not strictly upon etiologic agents such as microorganisms, but on function. Many of the endocrine disorders fall into this category. Hyperthyroidism and hypothyroidism are defined according to the level of functioning of the thyroid gland with respect to some defined normal level. As with the infectious diseases, this sort of knowledge is useful in that it provides a rational basis for treatment. When one understands how the thyroid functions, treatments for hyperfunction and hypofunction can be devised.

Syndromes

From a philosophical point of view, syndromes are the most interesting type of disease. Wulff (1976, p. 69) defines a "simple syndrome" as the simultaneous occurrence of a fixed combination of nosographic characters. However, more often it is the case that no single set of characters presents in every case that is said to represent a certain disease. In such cases, decisions must be made as to how many such characters it takes to say that the syndrome is present, and which characters count as part of the syndrome and which merely constitute background noise. Diseases then come to be defined according to what Wulff calls a "composite syndrome definition." He sees syndromes as preliminary definitions of disease awaiting a more fundamental explanation.

In fact, however, to see the syndrome as a single type of disease awaiting more precise definition may be misleading. As Uffe Juul Jensen (1984, pp. 63-73) shows, a great many diseases exhibit much variability;

the characteristics of standard cases are not shared by all cases of the disease. Attempts to define a disease by specifying its essence reflect only the state of the disease as defined at a particular time, perhaps only in the laboratory. In addition, actual disease kinds may themselves be changing, that is, undergoing evolution. Thus, not all syndromes may be diseases awaiting further explanation, but may instead represent variations in the broad spectrum of diseases. The following examples will serve to illustrate the complexity of the syndrome and the need for value judgments in the process of assembling groups of clinical observations into an entity called a syndrome.

Cushing's Syndrome

Cushing's Syndrome is an example of a disease that began as a simple description of a set of observations and gradually acquired an explanation that accounts for them. There were three stages in the evolution of Cushing's Syndrome as a clinical entity (Liddle and Shute, 1969, pp. 155-164). In the first period, Harvey Cushing merely noted the presence of basophil adenomas in the pituitary glands of several patients with obesity, weakness, osteoporosis, hypertension, purple abdominal striae, and a few other findings. The association of these findings was not well explained by Cushing. In the second stage, it was discovered that removal of hyperactive adrenal tissue cures the syndrome. Because cortisol is produced by the adrenal cortex, hypercortisolism was advanced as an explanation for the occurrence of the findings described by Cushing. In the third period, the causes of the hypercortisolism were determined. As it turns out, the same clinical syndrome explained by hypercortisolism can have distinctly different causes. The hypercortisolism can be caused by an adrenocorticotropic hormone (ACTH)-secreting tumor of the pituitary. This is the syndrome described by Cushing (although Cushing did not make the etiologic connection) and this specific entity has become known as Cushing's *Disease*. Cushing's *Syndrome*, however, has remained a "composite syndrome definition" and need not necessarily include a pituitary tumor. The set of people with Cushing's Disease is thus a subset of the set of people with Cushing's Syndrome. Cushing's Syndrome can also be caused by autonomous secretion of ACTH by non-pituitary tumors such as bronchogenic carcinoma or pancreatic carcinoma. However, ACTH need not be involved in Cushing's Syndrome at all. Excess cortisol can be produced as a result of primary adrenal hyperplasia or an adrenal tumor. In addition, Cushing's Syndrome can be caused iatrogenically by prolonged steroid use.

The upshot of all this is that a syndrome often has no precise definition, even clinically. There is certainly no nominal essence of Cushing's Syndrome, although cases may be described as bearing a Wittgensteinian family resemblance to each other. However, nosological values must be incorporated in decisions about what is to count among the characteristics of the family and what is not. These values will probably include scientific values, but, depending on the reasons for making the classification, could also include values such as aesthetic and political values. A number of people are obese and have osteoporosis and hypertension, but are not considered to have Cushing's Syndrome. Furthermore, some degree of hypercortisolism need not cause Cushing's Syndrome and the level necessary to cause the syndrome may not be the same in all individuals. So, there is no precise causal relationship. When syndromes have no precise stipulated definition, value judgments are inevitable in both definition of the syndrome and the diagnosis of particular cases.

Rheumatoid Arthritis

Rheumatoid arthritis illustrates some more of the complexities involved in defining a biomedical problem as a syndrome. In the interest of more precisely defining a syndrome, professional societies sometimes offer criteria for categorizing certain patients as having a particular syndrome. Such professionally defined criteria can also have strong political and economic implications. Thus, the criteria can incorporate nosological values from the political and economic realms of value.

In 1987, the American Rheumatism Association (ARA) adopted a new set of criteria for the definition of rheumatoid arthritis (Arnett, Edworthy, Bloch, et al., 1988, pp. 315-324). Previous criteria were complex, hard to apply, and yielded such categories as "definite," "probable," and "possible" rheumatoid arthritis. The clinical picture of rheumatoid arthritis includes many signs and symptoms also present in other conditions. In addition, not all patients with rheumatoid arthritis present the same set of findings. In proposing new criteria, the ARA was attempting to improve both the specificity of the diagnosis of rheumatoid arthritis, and also the simplicity of the criteria.

What the ARA proposed is a set of seven criteria: (1) morning stiffness; (2) arthritis of three or more joint areas; (3) arthritis of hand joints; (4) symmetric arthritis; (5) rheumatoid nodules; (6) serum rheumatoid factor; (7) radiographic changes. Each of these criteria is defined. For classification purposes, a patient is said to have rheumatoid arthritis if at least four of the seven criteria are satisfied.[9]

At first glance, it would appear that the conditions for having rheumatoid arthritis are simply being stipulated. Even if some of the criteria mentioned do represent empirically determined facts, the state of affairs consisting of having rheumatoid arthritis would simply be a matter of stipulating which and how many of the criteria are required in order to apply the stipulative definition. This would involve a value judgment about the relative importance of the criteria and which criteria were included or excluded. In fact, however, the criteria were chosen by comparing a group of patients with "known" rheumatoid arthritis, diagnosed by previous criteria, with patients free from symptoms and patients with other conditions that are easily confused with rheumatoid arthritis. After considering various combinations of proposed criteria, seven were chosen and it was decided that the presence of at least four most accurately separated people with rheumatoid arthritis from those who did not have rheumatoid arthritis. Thus, the criteria are not a pure stipulation of a syndrome definition, but an assertion that the stipulated criteria offer the best method for diagnosing the disease of rheumatoid arthritis.

This sort of syndrome definition depends on two important assumptions. First, it is assumed that there does exist a disease, rheumatoid arthritis, and that patients either have it or they do not. This claim is further supported by the authors' admission that this sort of syndrome diagnosis is merely a stopgap measure to be applied until the disease can be causally defined.

> Disease criteria which are descriptive reflect our current understanding of these disorders. Elucidation of specific pathogenetic mechanisms may at some point permit classification to be based directly on disease biology. However, these new criteria for RA will necessarily serve to improve understanding, classification, and comparability of patients with rheumatoid arthritis until other methods of achieving this purpose are available (Arnett, Edworthy, Bloch, et al., 1988, p. 323).

Second, the criteria themselves depend on the prior assignment of certain patients to the category of "known" rheumatoid arthritis for comparison with controls. This depends on a prior definition of the disease, which has been made by the experts on the committee along with the other investigators. Thus, no essence of rheumatoid arthritis is appealed to. The formulation of the categories is akin to the casuistic method of ethical analysis, in which new problem cases are resolved by comparing them with similar cases in which we are most sure of the answer.

The definition of rheumatoid arthritis as a syndrome, then, is not a mere stipulation, but an attempt to find a way to separate more confidently those cases that match what we have classically considered to be rheumatoid arthritis from those we are less sure about. But this process must surely make us recognize how much we construct syndromes with values as well as facts, for the decision about how sure we are that a patient has rheumatoid arthritis as well as the decision about who counts as an expert in this decision will reflect scientific and professional values as well as empirical facts.

Acquired Immunodeficiency Syndrome

Acquired Immunodeficiency Syndrome was first described in the United States in 1981 when several homosexual men were found to have *Pneumocystis carinii* pneumonia or Kaposi's sarcoma, diseases that are extremely rare in previously healthy individuals. Out of these facts was born a syndrome (first called "Acquired Immune Deficiency Syndrome," hence the acronym AIDS). The choice of the term "syndrome" suggests a clinical entity, but not an entity specified by an etiology. This reflects an uncertainty about the status of AIDS as a disease in itself.[10]

The Center for Disease Control (CDC) (1982, p. 508) first defined AIDS as "a disease, at least moderately predictive of a defect in cell-mediated immunity, occurring in a person with no known cause for diminished resistance to that disease."[11] In order to be classified as having AIDS, a patient might have one or more than one of these diseases. Thus, the new syndrome was described as consisting simply of the occurrence in a previously healthy individual of a disease usually found only in immunocompromised patients.

With the discovery of the linkage of AIDS with the Human Immunodeficiency Virus (HIV) in 1984, the definition became even more complex. The CDC definition has undergone regular revision since then. The 1987 definition has three parts. In the first part, AIDS was defined as an illness characterized by the presence of one or more of the "indicator" diseases, depending on the status of laboratory evidence for HIV infection. AIDS could be diagnosed without evidence of HIV infection if the patient had no other cause of immunodeficiency and a definitively diagnosed indicator disease. The second part of the definition concerns cases with evidence of HIV infection. With laboratory evidence for HIV infection, AIDS could be diagnosed even in the presence of other causes of immunodeficiency. In addition, certain of the indicator diseases needed to be diagnosed only presumptively, and not definitively. The third part of the definition allowed for the diagnosis of AIDS with laboratory evidence against HIV infection. In this circumstance, AIDS could be diagnosed only if both of two conditions obtain: (1) all other causes of immunodeficiency are ruled out, and (2) the patient has either *Pneumocystis carinii* pneumonia diagnosed definitively or any other definitively diagnosed indicator disease combined with a T-helper/ inducer (CD4) lymphocyte count of less than 400/mm^3 (Centers for Disease Control, 1987, pp. 45-65).

This sort of definition suggests a redundancy. One cannot extensionally define the class of AIDS patients without resorting to an intensional description of the classifications involved.[12] That is, one cannot say that a particular group of people has AIDS without knowing what we mean when we talk about "AIDS." In understanding what we mean by "AIDS" we seem to be appealing both to a disease in itself, and also to a collection of other indicator diseases. In defining AIDS both in terms of laboratory findings in themselves, and also in terms of its indicator diseases, we abandon the ideal of disjointness.

An idea of this sort has led John Lauritsen to insist that AIDS is not a disease. He maintains that AIDS is a "phoney construct," based on a tendency to reduce complex phenomena to a single cause. Given his assumption, based on the work of molecular biologist Peter Duesberg, that it is not the nature of retroviruses to cause disease, Lauritsen finds the notion of AIDS as a disease distinct from its indicator diseases to be absurd (Lauritsen, 1993, pp. 419-424).

Even if HIV had no etiological role in AIDS, Lauritsen's argument would not show that AIDS is not a syndrome, for the empirical occurrence of particular sorts of diseases in particular sorts of patients is the stuff of which syndromes are constructed. AIDS is undoubtedly constructed out of facts and values, but that does not make it a "phoney" disease any more than rheumatoid arthritis, which could also be construed as merely rheumatoid nodules, morning stiffness, arthritis of the hands, and so forth, all separate diseases of the symptom diagnosis type.

The CDC's 1993 definition of AIDS moves away from the classic syndrome definition and toward an etiologic definition. It expands the criteria for diagnosis of AIDS. The new definition retains the twenty-three indicator diseases of the previous (1987) definitions and adds three more: tuberculosis, recurrent pneumonia, and invasive cervical cancer. In addition, it expands the definition to include all HIV-infected persons with CD4+ T-lymphocyte counts of less than 200 cells per microliter, or a CD4+ percentage of less than 14 (Centers for Disease Control, 1992, pp. 1-19).

The expansion of the definition to include a diagnosis of AIDS in patients with a certain level of CD4+ T-lymphocytes even in the absence of any indicator diseases fundamentally alters the understanding of AIDS as a syndrome, and moves it toward a causally defined disease, similar to other viral infections. The nature of the infection in AIDS, however, is to cause an immune deficiency and therefore susceptibility to a number of the indicator diseases. Elements of the syndrome remain, however, insofar as indicator diseases still remain criteria for diagnosis on their own.

It should be kept in mind that the CDC has always maintained that AIDS definitions are surveillance definitions. The purpose of defining the syndrome is to assist in tracking the state of an epidemic, to project trends, and to assist in planning health care and prevention programs. The expansion of the definition reflects not only a commitment to moving from defining diseases as syndromes to causally defined diseases whenever possible, but also a conviction that identifying those with HIV infection and concomitant low levels of CD4+ T-lymphocytes will prove effective in preventing the onset of indicator diseases (Buehler and Ward, 1993, pp. 390-392). Thus, the definition of AIDS includes not only empirical facts, but also conceptual commitments (which include value judgments) about what constitutes the various indicator diseases and value commitments about prevention and treatment.

Sudden Infant Death Syndrome
The condition that has been named "sudden infant death syndrome" (SIDS) is another illustration of a condition without satisfactory etiological explanation that has come to be called a syndrome. Sudden and unexpected death of infants has been known since biblical times. In a dispute before King Solomon over a dead child, a woman asserts, "And this woman's son died in the night, because she lay on it" (1 Kings 3:19, Revised Standard Version). Thus, even in ancient times a link between

breathing difficulties and sudden infant death was made. Beginning in the nineteenth century, a variety of other etiological explanations were offered: for example, enlargement of the thymus causing interference with heart and lung function, laryngeal spasm, bacterial infection, and aspiration of vomit. Nonetheless, accidental suffocation by "overlaying," or when the infant slept alone, by bedclothes, remained the most common explanation (Limerick, 1992, pp. 3-4).

In recent years, extensive research on SIDS has been carried out. Typical findings at autopsy include a well-developed and well-nourished body, a small amount of fluid in the nares, cyanosis of lips and nail beds, petechiae in the thymus, pleura and pericardium, "subacute" inflammation in the upper respiratory tract, foci of fibrinous necrosis in the larynx, full expansion of the lungs with pulmonary congestion and edema, liquid blood in the heart, prominence of all lymphoid structures, and an empty urinary bladder (Valdes-Dapena, 1992, pp. 702-704). Certain risk factors, such as prematurity, low birth weight, male sex, race, climate and season, and prone sleeping position have been suggested (Hoffman and Hillman, 1992, pp. 717-737). Such findings have led to much speculation about the cause of SIDS, but none of these findings is significant enough to explain the death of the infant.

Despite all this uncertainty, the fact is that significant numbers of infants between one month and one year of age die suddenly and unexpectedly. This has led to the construction of a most peculiar syndrome. The currently accepted definition of SIDS, as adopted by the Second International Conference on Causes of Sudden Death in Infants, is:

> The sudden death of any infant or young child, which is unexpected by history, and in which a thorough post-mortem examination fails to demonstrate an adequate cause for death (Beckwith, 1969, p. 18).

Criteria for minimal acceptable port-mortem investigation are given. What is most interesting about this definition is that it includes none of the grouping of facts that are typical of syndrome definitions. The definition of SIDS is purely a definition of exclusion. The only fact is that there is an infant who was normal when put to bed and later found dead. This leaves ample room for value judgments to come into play in constructing SIDS. For example, deciding whether or not a post-mortem investigation is adequate can involve a stipulation by experts, but the notion of adequacy will inevitably involve scientific and professional values. Thus, to say that autopsy failed to show a cause of death involves not only facts about the condition of various organs, but also, for example, whether or not the amount of pulmonary edema present was sufficient to cause death in the particular case.

A disease without physical findings is an anomaly. This has led some to suggest that SIDS is a "nonviable concept" and should not be considered to be a disease distinct from other recognized diseases.[13] Another possibility would be to admit that SIDS is neither a definition of a disease nor a cause of death, but rather is simply a conclusion about a pathologist's inability to find a cause of death. Thus, SIDS would not be a category of disease, but a category of death (Freed et al., 1994, pp. 969).

Another suggestion is to change the definition of SIDS. An expert panel of the National Institute of Child Health and Human Development has recently suggested the following:

> The sudden death of an infant under one year of age which remains unexplained after a thorough case investigation, including performance of a complete autopsy, examination of the death scene, and review of the clinical history (Willinger, James and Catz, 1991, p. 681).

This definition, however, expands the definition of a disease beyond the expertise of most physicians in its demand for examination of the death scene. This is more akin to police work or at least to epidemiological investigation of occurrences such as outbreaks of infectious diseases. Even in such cases, however, no definition of disease depends upon the investigation, but merely a judgment of how individuals became affected by already well-defined biomedical problems.

A recent international conference on SIDS could reach no consensus on a new definition of SIDS and decided to retain the 1969 definition. The primary issue in dispute was what constitutes an adequate cause of death. This is a problem because, as we have seen, a number of rather constant findings are seen at autopsies of infants who have died suddenly. French pathologists disagree that all such findings are insignificant. They propose a classification according to main post-mortem findings, such as SIDS with myocarditis (Rambaud, Guillenminault, and Campbell, 1994, p. 1439). Others, however, continue to insist that such findings are insufficient explanation for the death of the infant, and maintain that changing the definition of SIDS in this way would tend to obscure further research on the etiology of SIDS (Mitchell, et al, 1994, p. 607).

Several problems exist with the use of the term "SIDS." Epidemiologists have tended to treat SIDS as a pathogenic entity. SIDS, in fact, is given the code 798.0 in ICD-9. However, many feel that SIDS merely represents a "common final pathway" of several disease processes. To see SIDS as a single entity would be to obscure those processes (Limerick, 1992, p. 5).

SIDS is said to account for approximately forty percent of post-neonatal mortality in the United States and similar developed countries (Hoffman and Hillman, 1992, p. 717). As a single entity this condition assumes great importance, and may dictate that great efforts at fund raising and research be carried out. Thus, there is much at stake in the definition of disease.

Disease classification is primarily a useful tool for physicians in dealing with individuals coping with illness in formulating treatments and offering prognoses. However, SIDS shares little in common with other syndromes or any other variety of disease. It is unclear whether looking at sudden unexplained death of infants as a single disease is useful in this way.

VI. CONCLUSIONS

So far I have argued that value judgments are necessarily incorporated into the entities we see as diseases at three levels. First, any scientific fact is value-laden. Foundational values in many realms influence what we take to be a fact. Scientific facts, in particular, are the product of a particular sociocultural milieu with its sociocultural, scientific, and professional values. Second, to call some state of affairs a disease, even if that state of affairs consists primarily of scientific facts, another layer of values will be involved. "Disease" usually connotes disvalue, and so the concept of disease includes what I have called conceptual values. Diseases may disrupt the pursuit of many sorts of values, sociocultural, economic, and aesthetic, and religious. Third, to classify certain types of diseases as being like others, that is, to construct a nosology, requires yet another layer of values, the nosological values discussed in this chapter.

If this is so, when physicians make diagnoses, they are dealing not only with bare facts, but also with three types of values, which are embedded in the biomedical problems they are trying to diagnose. In the next two chapters, I will argue that a fourth level of values necessarily comes into play in the diagnostic process itself. It is this fourth level where moral values become important.

PART THREE

DIAGNOSIS

CHAPTER 6

THE ELEMENTS OF DIAGNOSIS

I. INTRODUCTION

So far, we have been considering the concept of disease and the classification of diseases largely from a metaphysical standpoint. I have argued that diseases are socially constructed, but that this does not mean diseases are merely made up to suit our present fancy. Diseases, and biomedical problems in general, are constructed from facts and from values. We have looked at various ontological conceptions of disease and various nosologies. We have seen that foundational values are essential parts of the facts that constitute them, conceptual values are essential parts of our understanding of the concept of disease, and nosological values are necessary to classify various collections of facts as different kinds of diseases. Hence, in any biomedical problem there is an intimate connection between fact and value.

Our emphasis now shifts from metaphysics to epistemology as we consider how we come to know disease through the process of diagnosis. My task is to examine the ways that values are incorporated into the diagnostic process. I will call these *diagnostic* values. Some diagnostic values are of a similar nature to the values of the previous three levels. That is, the process of diagnosis will involve aesthetic, scientific, professional, and sociocultural values akin to those that are involved in formulating facts and the concept of disease and in the construction of nosologies. However, we will see that diagnostic values importantly include moral values in a way that we have not yet encountered. This is so because the nature of the diagnostic process is that it involves a living human being who is the subject of investigation. The physician-patient encounter in the diagnostic process is a relationship filled with moral import, for decisions about information to be divulged, physical manipulations to be made, and laboratory tests carrying a degree of risk to the patient are all involved in the process of diagnosis.

The process of diagnosis is complex. However, "diagnostics," the general theory of diagnosis (Laor and Agassi, 1990, p. 2) can be described quite simply as consisting of two steps: the gathering of information, and the attempt to see a pattern in terms of already established classifications (Clouser, 1985, p. 38). As we will see, these two steps are intertwined in complex ways, and so should not be taken as a temporal or even a logical sequence. Nonetheless, it is convenient to conceptually separate them for purposes of analysis, if only to show the complexity of their interaction. In this chapter, I will consider the first of these two steps and show how fact and value interact in the information gathering process. The second step will be the subject of the next chapter.

The information gathering stage of the diagnostic process has three major elements: obtaining a history of the patient's illness, performing a physical examination, and performing laboratory and other diagnostic tests. Information gathering seeks facts, which are already value-laden, and adds additional diagnostic values in the process.

II. THE HISTORY OF THE ILLNESS

In most cases, when a patient comes to a physician with a complaint of feeling ill, the physician begins the information gathering process by "taking a history." That is, the physician guides the patient in telling the story of the patient's experience of illness. Alvan Feinstein (1967, p. 299) sees history taking as far more complex than the techniques of physical examination and laboratory tests; he calls history-taking "the most clinically sophisticated procedure of medicine."

Part of the complexity arises from the fact that the history-taking process is essentially an attempt to communicate information about experiences that are largely private (e.g., pain) from one human being to another. The ideal purpose of the history is to record as exactly as possible the patient's experience of illness. However, as the history is recorded by the physician, it must necessarily be interpreted. Feinstein's major concern is that physicians do not introduce bias into histories by the use of ambiguous and imprecise language (e.g., heartburn). If Feinstein means *conscious* distortion of what the patient presents, I agree with him. The physician's tendency is to want to hear the patient's descriptions of the illness in terms of already established categories of disease. A physician may ask inappropriate or leading questions and even subconsciously manipulate a patient into giving a history that is not quite true to the patient's experience, but fits some established pattern (Feinstein, 1967, p. 309). Because of at least partial incommensurability between the worldviews of physician and patient, however, it is impossible to record a patient's experience without any sort of bias.

A patient's symptoms may become known to a physician in three different ways: they may be spontaneously mentioned, they may be elicited by routine questioning, or they may be revealed by specific questioning when a particular diagnosis is suspected by the physician (Wulff, 1976, p. 18). The contribution of the physician is evident when information is gleaned as a result of questioning. However, even when symptoms are spontaneously mentioned, they are almost never articulated in a way that will be accepted by the physician without further probing. The fact that the history-taking process is an interactive conversation means that the history that is finally recorded will contain significant contributions of the physician. Many possible accounts of the patient's symptoms will be left out entirely. Medical histories, insofar as they are interpretations of experience, reflect conceptual and value judgments by both physician and patient about what terms to use and what parts of the patient's myriad experiences of aches and pains are significant enough to be recorded in the history.

Furthermore, the act of history taking is a transformative act. It changes both the sick person and the person's sickness. Arthur Kleinman (1988, pp. 130-131) writes:

> The recording of a case in the medical record, a seemingly innocuous means of description, is in fact a profound, ritual act of transformation through which illness is made over into disease, person becomes patient, and professional values are transferred from the practitioner to the "case." Through this act of writing up a patient account, the practitioner turns the sick person as *subject* into an *object* first of professional inquiry and eventually of manipulation.

The professional values of which Kleinman speaks are an important category of diagnostic values. Before I discuss these, however, I want to consider the values that the patient contributes to what becomes the patient's history.

The Patient's Contribution to the History

Patients bring value systems to their encounters with physicians. Some patients may find certain of their experiences unimportant and fail to report them to the physician. Others may flood the physician with reports of every ache and pain they have experienced in the course of a day and thereby force the physician to make judgments not only about what is relevant and what is not, but also about the validity of the patient's own experience.

Patients may not understand questions they are asked by the physician, and give information that is misleading to the physician. In one study, more than a third of men who were asked, "Are you circumcised?" gave the "wrong" answer (Schwartz and Griffin, 1986, p. 30). Presumably, these men were aware of the relevant facts about their physical state, and simply did not understand the meaning of "circumcision" as used by the physicians who asked the question. The facts here seem clear. When the facts addressed in the conversation are laden with more controversial and covert conceptual and value commitments, however, there is even more room for misunderstanding. Significant but covert diagnostic values in the history may go unnoticed by the physician.

Problems arise in communication because the physician's task in taking a history is to obtain information that will explain the experience of the patient in a reductionistic manner, in terms of malfunctions of organ systems, individual cells, and so on down to the molecular level. Even in this biomedical mechanical model, however, there may be several possible accounts. Patients, however, are generally unaware of these complexities. They are not likely to relate their experiences in ways that are directly translatable into the biomedical mechanical model. While most patients would see this translation as essential for their well-being, many would see it as insufficient. Patients seek healing, and the experience of healing often comes as much from a feeling of being understood as from a relief from physical pain. If this value, brought by the patient to the physician-patient encounter, goes unrecognized by the physician, an important dimension of the patient's story is missed.

The patient's explanation of an illness may be said to exist at a different hierarchical level, the level of first-hand experience. This may not be directly translatable into medical terminology. However, even at the level of first-hand experience, the patient also has many possible conceptual and evaluative choices that must be made in reporting a history. With all of these possibilities to consider, the complexity of trying to relate the levels of patient and physician is daunting (Blois, 1988, pp. 847-851).

The Physician's Contribution to the History

The most significant contributions of the physician to the medical history are those that result from specific questions about hypotheses being entertained by the physician. As the physician takes a history from a patient, the physician is attempting to understand the patient's illness in terms of already established nosologies. The patient's report of illness is translated by the physician into the language of disease, which is an explanatory language in the particular worldview of scientific medicine (Engelhardt, 1981, pp. 301-302).

Expectations on the part of the physician are likely to influence the direction of the conversation during the taking of the history. Not only do physicians think in terms of preestablished nosologies, but they are also likely to favor moving the history taking process toward making diagnoses for which effective treatments are available. Physicians will pursue different aspects of the patient's illness depending on various evaluative decisions about the consequences of their diagnoses (Engelhardt, 1980, p. 45).

A Narrative Approach to the History

One of the major themes I have tried to stress throughout this book is that although a rigorous scientific method applied to the healing arts has resulted in the greatest advances in the history of medicine, science does not tell the whole story in the explanation of an illness. This is perhaps most evident in the comparison of the patient's and the physician's versions of the medical history. Some recent work in the medical humanities has looked at hermeneutical problems in medicine in light of critical literary theory. This approach can help us to understand the interaction of scientific and non-scientific elements in the medical history.

In this narrative approach, a patient may be understood in a way that is analogous to the way we understand a literary text. A text is a group of signs, which as a whole takes on meaning through interpretation. The text may be interpreted on four levels: the object of interpretation, the mode of interpretation, the life-affecting activity (praxis) following from interpretation, and the change of life-world brought about through the interpretation (Daniel, 1986, pp. 199-202). In terms of medicine, the object is the patient, and the mode of interpretation is diagnosis, which leads to the praxis of therapy and the resulting change in life-world of healing.

In diagnosis, the primary text is the patient, but this text can only be read in terms of two secondary texts. The first is oral and is the patient's history. The second is written and is the physician's documentation of the clinical interaction. The physician must interpret both the patient and the patient's story (Daniel, 1986, p. 202).

When diagnosis is understood in terms of this literary analogy, the hermeneutic circle comes into play. That is, in order to interpret a text, one must already be familiar with the parts of the text such as the vocabulary and grammar, and have some idea what the text as a whole might mean. In listening to a patient's history of an illness, the physician will tend to postulate meaning in terms of preliminary diagnoses, refining these first guesses by asking questions to rule out particular guesses (Daniel, 1986, p. 205). The physician's own sociocultural, professional, scientific, and aesthetic values come into play in choosing which strands of the patient's story to follow and which to ignore, and in choosing what to retain as significant and what to reject as background noise. Further value judgments must be made in considering what parts of the story deserve to be recorded in the medical record. This decision has overt ethical implications because the story becomes a written record that formally labels a patient as fitting the sick role in a particular way. This carries significant social implications for the way the patient may be treated in the future. Labeling someone as having AIDS, for example, can fundamentally alter social relationships and can lead to unjustified employment discrimination.

Partial Incommensurability of Narratives

What results from the physician-patient history-taking encounter are two stories: the patient's and the physician's. While the two are related, they are different stories. They have been constructed from different points of view and with different motivations. Kathryn Montgomery Hunter (1991, p. 13) describes the two stories:

> The first, the patient's story, is the original motivating account that the person who is ill (or family or friends) brings to the physician; the second is the medical account constructed by the physician from selected, augmented parts of the patient's story and from the signs of

illness in the body. The first concerns the effects of illness in a life, the second the identification and treatment of a disease.

The creation of the second, medical, story is the physician's immediate goal in the patient-physician diagnostic encounter. The medical story, according to Hunter, is a reinterpretation of the patient's story in terms of medical science. By "medical story," Hunter apparently means the story told by the medical professional. There may be more than one medical interpretation of the patient's story, however, and hence more than one possible medical story. Which one is chosen by the physician will reflect that particular physician's conceptual and value commitments.

While the reinterpretation of the patient's story is important to both physician and patient, Hunter argues that it ought not to alter the priority of the patient's experience. She says:

> The physician's narrative, although it takes on a life and meaning of its own, is secondary, derivative. The patient's account of illness remains the fundamental fact in clinical medicine. The events of illness do not exist for the physician's narrative, even though the physician's narrative, as the process and record of the diagnostic circle, yields us their medical meaning (Hunter, 1991, pp. 14-15).

Knowing medical facts about a patient, then, is rarely a matter of unilateral discovery on the physician's part, for the physician depends on the patient's story, which is itself an interpretation of experience, to provide the data that figure in the generation of the facts.

In Hunter's view, the patient's account of illness and the medical version of that account are fundamentally different narratives. The medical narrative is not simply a translation of the patient's story into medical terminology. No medical narrative is a life story. Hunter thinks that the two accounts are incommensurable in the sense that neither version can be exactly translated into the terms of the other. What is needed, if the patient is to experience healing, is for the physician to return the medical narrative to the patient in such a way that the patient is able to integrate the medical ending into the patient's own non-medical life story (Hunter, 1991, pp. 123-147).

If the patient's and the physician's narratives of an illness are truly incommensurable, in the radical sense associated with Thomas Kuhn's earlier writing, it would seem that there is no way to translate one into the other. One might well ask why a physician would bother to elicit a history from the patient if the patient's narrative is radically incommensurable with the medical narrative the physician is attempting to write. Relying wholly on physical examination and laboratory testing, both of which lie in the realm of the medical worldview, would obviate the need to attempt any such translation.

Physicians do, however, place great importance on the history of the illness as reported by the patient. While incommensurability may play some role in the difference between the two sorts of narratives, the incommensurability cannot be such that it makes communication between patient and physician impossible. Robert Veatch and I have argued that differences in worldviews of patient and physician raise difficulties in the physician-patient relationship with respect to understanding medical science and recommending treatments. However, because the worldviews of patient and physician are only *partially* incommensurable, meaningful communication can take place (Veatch and Stempsey, 1995, pp. 253-269). If worldviews are only partially incommensurable with respect to treatment choices, it is reasonable to think that the narratives in the diagnostic process are only partially incommensurable as well.

First, physicians may be able to take a non-medical view of a patient's illness. Even though their training will inevitably change the way they view illness, physicians still may be able to set this training aside to some extent and appreciate a patient's experience of illness. Second, patients, with some study and experience, may be able to experience illness from a medical point of view. A person with diabetes mellitus may learn to correlate a particular subjective experience with a too-low glucose level, thus interpreting the subjective experience in medical terms. The difference in worldview of patient and physician, then, would lie more in the relative importance given to one or other version of the illness/disease narrative.

We might consider one other aspect of incommensurability, an aspect suggested in our previous discussion of ontological conceptions of disease. A disease that has a history is a disease conceived temporally. Other conceptions of the same disease, especially anatomic conceptions, may well emphasize the spatial, as did Foucault's.[1] This incommensurability is not between the worldviews of patient and physician, but in the very concept of disease. It is probably not helpful to try to conceive of disease as either exclusively temporal or exclusively spatial. However, one of these views is likely to be predominant in a particular disease. Whether the physician will emphasize the temporal aspects of disease or the spatial aspects will depend to a large extent on a value judgment about the usefulness of analyzing the particular case in terms of time versus space. Usefulness depends on the end that is sought in making the diagnosis, and such ends depend on what one values.

III. THE PHYSICAL EXAMINATION

As we saw in the last chapter, the development of instruments such as the stethoscope enabled physicians to observe the workings of the bodily organs during life, and this launched the anatomico-clinical understanding of disease. This model of disease provides the conceptual basis for understanding the value of the physical examination. Through examination of the body, the physician claims knowledge of the workings of the inner organs, workings that remain hidden to ordinary vision.

The Need for Interpretation of Physical Findings

The kind of incommensurability we found in the taking of the patient's history should not present a problem in the physical examination. In the physical examination, the physician observes the body of the patient firsthand and does not need to rely on the patient's experience to any great extent. The patient may lead the physician toward examining particular areas of the body, and it may be necessary to interpret whether a patient is experiencing tenderness on palpation and the like. However, interpretation of the patient's experience in the physical examination does not play the significant role that it does in obtaining the history. In the physical examination, the physician simply makes observations and records them directly in the medical narrative.

The naïve realist might even take the physical examination to be a simple matter of discovering facts, subject only to the epistemic uncertainties inherent in human observation. To be sure, several physicians who examine a single patient may report different findings. These differences, in such a realist account, simply reflect the fact that at least some of the observers, and possibly all of them, are mistaken in their reporting of the facts.

Just because interpretation across the worldviews of patient and physician is unnecessary in physical examination, however, does not mean that the physical examination provides the physician with bare, uninterpreted facts. The facts of the physical examination reflect the conceptual, psychological, and moral commitments on which they are built.

Conceptual Issues

The Subjectivity of Sensations

At least part of the problem of inter-observer variation lies in the nature of the experience of sense data in the physical examination. As Wulff (1976, p. 23) points out, if one physician hears a few crepitant râles and another does not, there is no means, given only the methods of physical examination, of telling who is right. To resolve disputes about observed physical findings, physicians usually seek some independent means of confirmation, most often a diagnostic test that is believed to correlate with the physical finding in question. Without some such agreed-upon standard, disputes are usually settled by appeals to authority, using the judgment of a respected senior diagnostician as the factual standard. This, however, is no real solution to the problem. (See the section on the influence of authority in *Psychological Factors*, below.)

As Joseph Agassi (1976, p. 310) argues, we cannot avoid reading meaning into our perceptions. Furthermore, we do not know how much of our observation of what we call facts is perception and how much is theory. To say this is to deny John Locke's empiricist sensationalism, the view that all our ideas are built up from individual sense perceptions. It is not, however, to say that our perceptions are properly dependent on any random meaning we might choose to give them. It must be kept in mind that even a common experience such as causality cannot be explained satisfactorily to a majority of philosophers. Agassi pragmatically suggests that although we cannot explain causality to the liking of David Hume's disciples, we cannot avoid using causality to explain many of our everyday experiences. We cannot avoid seeing even the plainest of facts without resort to such hidden explanations. The more familiar we are with instances of a particular fact, the more we tend to see similar facts in terms of the same hidden mechanisms. Thus, our observations are never absolutely provable, but are always more or less conjectural, dependent on theory, and open to future refutation (Agassi, 1976, pp. 310-312).

Objectivity

Feinstein deplores the "defective science" in some examinations performed by physicians. He maintains that objectivity is destroyed when an examiner is already aware of the findings of others (Feinstein, 1967, p. 310). This makes some sense. The suggestion of another can make us see ambiguous data in the manner that has been suggested. However, uninfluenced observation cannot be construed as a way to overcome "defective science" if uninfluenced observation is considered a privileged route to uninterpreted facts. When skilled observers independently reach the same conclusions regarding physical findings, we have strong evidence that the findings are objectively correct. Nonetheless, it must be remembered that any report of physical findings carries several layers of conceptual commitments along with the value choices that have gone into those conceptual commitments. Thus, the objectivity Feinstein seeks cannot be the objectivity of the naïve realist. Value-dependent realism maintains the possibility of objectivity in the context of conceptual and value commitments.

Concept-Laden Descriptions

Certain observations carry with them implicit interpretive judgments. Koplik's spots in the buccal mucosa are pathognomonic for measles; when a physician observes Koplik's spots, the diagnosis of measles is certain. The problem lies in determining whether what one observes— small, bluish-white specks in an irregularly shaped background of red— are really Koplik's spots. In claiming that they are, the physician is moving from a pure observation language to a language constructed for the purpose of identifying a disease already framed in terms of a particular nosology (Engelhardt, 1981, pp. 305-306). The description of spots in the observation language carries with it a certain commitment to concepts and foundational values. The general agreement in the pertinent society of observers on these conceptual and value commitments enables the observers to take the presence of spots of a particular description as a fact. To claim that they are Koplik's spots, however, entails an association with measles. This also necessitates the acceptance of the nosological values that have gone into classifying a certain set of observations as the particular disease, measles. Furthermore, to call measles a disease entails the inclusion of conceptual values.

The Problem of the Standard

The choice of which method is to serve as the "gold standard" to decide which physical findings are "real" will ultimately involve a value judgment. Diagnostic tests such as x-rays will usually be used when anatomic structures are in question. In other cases, when a physiological function is in question, particular laboratory tests will be used. Such standards may depend on a particular metaphysical view of disease, but more often will simply reflect the more pragmatic value of what will be useful in making prognoses and offering treatment. Diagnosis is carried out for some goal, and as we found in our discussion of function, to specify goals demands recourse to values. Deciding on the standard that will be most useful involves the balancing of diagnostic values based on the end that is sought in the diagnostic encounter.

Psychological Factors

The Effect of Assumed Probabilities

Empirical studies in the psychology of perception have shown that observers are much more likely to see what they consider to be highly probable than what they consider to be rare.[2] Oliver Sacks recounts his experience of meeting Witty Ticcy Ray, a man with Tourette's Syndrome. The following day, walking on the streets of New York, Sacks noticed three other people with Tourette's Syndrome within an hour. This astonished him, for he believed Tourette's Syndrome to be exceedingly rare (Sacks, 1990, p. 94). Sacks questioned his belief about the incidence of Tourette's Syndrome; others might have questioned whether the three people he saw actually had Tourette's Syndrome. Reporting the facts about who was on the street that day will depend on the relative value one puts on various beliefs about one's knowledge when those beliefs come into conflict.

Distortions Due to Expectations

Daniel Kahneman and Amos Tversky (1982, p. 144) aptly describe the physician's plight with respect to obtaining information from a physical examination.

> At all levels of biological complexity there is uncertainty about the significance of signs of stimuli and about the possible consequences of actions. At all levels, action must be taken before the uncertainty is resolved, and a proper balance must be achieved between a high level of specific readiness for events that are most likely to occur and a general ability to respond appropriately when the unexpected happens.

Two sorts of expectations play a role in what we actually perceive. Active expectations draw on a limited span of attention, as when one is searching for some particular thing. Passive expectations, on the other hand, are automatic, more like dispositions than activities. Some passive expectations are relatively permanent and reflect the categories and assumptions that we use in our interpretations of our perceptions. Such permanent passive expectancies help us in interpreting ambiguous stimuli. Other passive expectancies are more temporary, mediating the effects of context on particular perceptions. Kahneman and Tversky argue that, in general, passive expectancies are more important determiners of our perceptions than active ones. We have a passive expectancy for rooms and windows to be rectangular. When ambiguous stimuli are presented, we tend to see rooms and windows as rectangular even if this means altering other perceptions in ways that we would not expect. The outcome of this is a tendency to perceive even familiar people as changing size as they walk along the wall of a distorted room (Kahneman and Tversky, 1982, p. 147).

It might be expected, then, that physicians will tend to interpret ambiguous stimuli in physical examinations according to some passive expectancies. These are likely to be influenced by the categories and assumptions of the medicine they have learned, and this, as we have seen, is embedded with conceptual choices and value judgments on several levels. This phenomenon can lead to distortion and overinterpretation, as when fingernail clubbing is reported in elderly patients who simply have curved fingernails (Bradley, 1993, p. 52). However, because all observations are influenced to some extent by such expectations, we need not say that all such influence leads to distortion or bias. The more clearly we can describe these influences, the better we will be able to recognize the conceptual and value judgments built into our observations.

The Influence of Authority

Two sorts of authority can influence the perception of physical findings. One is the presumed epistemic priority of technological means of "seeing." The other is the influence of a physician who is perceived as an authority on physical diagnosis.

Henrik Wulff (1976, p. 22) recounts stories in which clinicians "miss" diagnoses. A physician listens to a patient's chest but finds nothing abnormal. A routine chest x-ray, however, shows a density in the left lung. With this knowledge, the physician listens again and this time hears dullness and bronchial respiration on the left side. Another patient is examined by two physicians who report no cardiac abnormalities. A third physician, however, hears a systolic murmur, which is then confirmed by a phonocardiogram. All three physicians then are able to hear the murmur without difficulty.

The tendency to trust one who is perceived to be a higher authority has even inspired the construction of a new disease, "The Emperor's Clothes Syndrome" (Gross, 1971, p. 863). This syndrome usually affects those in middle levels of medical training. In a typical case, the chief of service, while making rounds in the cardiac care unit, examines a patient and hears a particular heart sound that was undetected by previous examiners. The syndrome is made manifest when the senior resident says, "I hear it" and others, in rapid succession, also hear what they previously failed to hear.

Physical findings must be perceived if they are to be considered facts, and perception has a significant subjective component. Because this is so widely recognized, physicians are usually very strict about seeking corroboration for their observations. As Wulff (1976, pp. 22-23) points out, the accuracy of a physical finding can only be determined if the finding can be confirmed or disproved by some other method. The authority of one who is deemed experienced in physical examination is one method, but not the one trusted by most physicians. Diagnostic tests are usually favored as more objective.

All this suggests that the observations made by the physician in the physical examination are not observations of pure fact, but observations made in light of particular conceptual commitments. These conceptual commitments include the values embedded in the particular nosology with which the physician approaches the patient, and the particular epistemic values established for the practice of physical diagnosis by the medical community at large. The observations of a physical examination are also made for a purpose, which carries its own particular values. The primary purpose of diagnosis is to establish a basis for rational treatment (and what counts as rational treatment carries value judgments in itself).

Moral Aspects

The physical examination introduces overtly moral, or ethical, concerns as well. These moral values are fairly evident, and would be recognized even by those who hold that facts about disease are value-free. In examining a patient, several sources of potential discomfort for the patient arise. The examiner can cause pain in manipulating the patient's body. Carelessness in providing proper covering for the body of the patient can cause embarrassment. When such values have to be sacrificed, or mitigated to some extent, in order to obtain desired information from the physical examination, it becomes evident that moral values have come into play in the attempt to obtain the facts of the case. How much discomfort should a patient be forced to endure in an attempt to learn the facts that physical examination might disclose? This is a moral question. The moral values that come into play in answering such questions are among the most important of the diagnostic values.

IV. DIAGNOSTIC TESTS

The Purpose of Diagnostic Testing

Physicians have increasingly turned to various forms of diagnostic testing in an attempt to mitigate the problems of inter-observer variability in history taking and physical examination. Some such tests extend the physician's ability to observe various parts of the patient. Light and electron microscopy offer the physician enhanced views of the patient's internal organs, blood and body fluids. X-ray examinations, especially those enhanced by the use of contrast media, allow the physician to infer the anatomy of internal organs. Computerized tomography, magnetic resonance imaging, and positron emission tomography have furthered the extension of the medical gaze to within the living body, and have even allowed the observation of physiological processes as well as anatomic structures.

Other laboratory tests give information about disease processes. Quantification of various constituents of the body, such as tissue, blood and other fluids, allows the physician to make inferences about the physiological functioning of the body.

A third use of diagnostic tests is to identify the various etiologic causes of disease. The bacteriological culture of wounds is a good example of this kind of test. Toxicology screening tests also can help identify etiologic agents responsible for cases of poisoning. Such tests are often crucial for determining proper treatment.

Finally, diagnostic tests can detect various markers associated with increased risk of disease in people who are presently healthy. Such tests will assume increasing importance, as the genetic basis for many diseases becomes better understood. The ways in which such information may be used carry important social and ethical implications.[3]

All diagnostic tests incorporate value judgments. Furthermore, the results of all such testing must be interpreted if they are to be of any use. Interpretation will further incorporate value judgments into the value-laden facts they present. In this section, I will examine several aspects of laboratory testing to show how values are brought into them.

Choice of Tests

Whether a test ought to be done in a particular case is a normative question. Once it is decided that a test ought to be done, the choice of which particular test to do involves further normative elements. These questions can be assessed by a complex cost-benefit analysis, which, in the account of Henrik Wulff (1976, pp. 109-110), takes into account five factors. The first is the value of the test for establishing or excluding the diagnosis. Wulff's language here is overtly evaluative. He rightly claims that a value judgment must be made in deciding how well a test fulfills its purpose as an aid for establishing a diagnosis. This presumes that one already has made a hypothesis about possible diagnoses. Hence, foundational, conceptual, and nosological values are already presumed before the value judgment about which Wulff speaks is brought into play.

The second factor is the consequences for the patient of establishing or excluding the diagnosis. Some diagnoses may have little importance for the patient. Other diagnoses, however, may put patients into categories that have profound implications for the ways in which they will be treated in society. This introduces an important moral component into making a diagnosis.

Reducing uncertainty may lie behind requests for some sorts of tests. Some individuals want to know whether they are at risk for certain diseases. For example, genes associated with breast cancer have been identified. Research has shown that women who are interested in reducing uncertainty and protecting future health will want the test (Lerman and Croyle, 1994, pp. 609-616). The presence of the gene is not a certain prediction that the disease will develop. Likewise, the absence of the gene is not a certain prediction that the disease will not develop, for other factors have a role in producing breast cancer. However, the women studied apparently would value knowing whether the statistical likelihood of their getting breast cancer is increased.

Wulff's third factor is the discomfort and risk to the patient. While the physical examination carries primarily a risk of psychological harm, various diagnostic tests can expose the patient to what the patient might consider a significant risk of bodily harm. Risk-benefit assessments must consider the possibility of harms of various sorts to the patient in relation to the information that can be obtained from the diagnostic test. The moral values involved here are most important. When a disease considered serious (from some value perspective) is treatable and can be diagnosed with a level of certainty considered adequate (from some value perspective) by a blood test considered simple (from some value perspective), the value judgment to do the test can easily be made, assuming the patient wants the treatment. Likewise, if a procedure considered very risky (from some value perspective), something like a brain biopsy, is unlikely to yield information that is significant (from some value perspective), or can be used only to confirm a particular disease for which no treatment is possible, most patients would conclude that such a test is not justified. Most cases, however, fall somewhere in between, and will require complex balancing of values rather than more simple value judgments of these extreme cases. At this stage, it should be noted, a patient's values and a physician's values may not match.

The fourth factor is economic cost. The cost of diagnostic testing includes wages to medical personnel, cost of instruments, and possibly the cost to the patient of hospitalization and loss of income. This involves value judgments about what potential information is worth from a financial standpoint. In the past, such matters have not been of much concern to physicians, but with the advent of managed care, physicians (and patients) are being forced to consider costs. Value conflicts between physician judgment and corporate policy as well as between physician and patient are possible. The basic issue is how much money a new fact is worth.

Wulff's fifth factor is what he calls "data pollution." If enough tests are performed, conflicting facts are bound to surface. This tends to confuse the diagnostic picture rather than to clarify it. The likely outcome is that even more diagnostic tests will be ordered in an attempt to clear up the picture. This may help, but it may also cloud the picture even further. As more and more measurements become possible, the number of ascertainable facts about a patient's anatomy and physiology increases. Not all such facts will be useful. Whether a fact is a useful fact will depend on the end for which it will be used, and this end will depend on particular values. Again, the values of physician and patient may differ on whether more information is desirable.

Diagnostic values from several realms are involved in these five factors. First of all, notice that speaking of test choice in terms of a risk-benefit analysis brings values into play. Benefits are things that are valued, and risks involve the increased likelihood of things that are disvalued. The values and disvalues of testing are of several different kinds. Deciding how well a test fulfills its purposes involves scientific values. When we examine how the test results will affect the patient, we enter into the sociocultural realm of value. Recognizing that some tests put a patient at risk and cause discomfort is an indication that we are dealing with moral values. A fundamental ethical demand is that we do not cause harm to people without due cause. To take economic costs into consideration when deciding to do a test is more than to establish the facts about the cost of a test; it is to make a value judgment in the realm of economic values. Is the information that might be gained from the test worth the cost? Finally, to call an excess of information "data pollution" carries an inherent value judgment. Pollution is not a good thing. The values here are epistemic and scientific.

Interpretation of Test Results

Once tests are chosen and carried out, the results must be interpreted. The value-laden facts that are the result of testing acquire additional diagnostic values in the process of interpretation.

The Problem of the Data Glut

Stanley Reiser (1978, p. 195) traces the development of the use of technology for diagnosis. He argues that physicians' habits of relying on technological devices for diagnosis has produced a "glut of medical data," which requires the further development of new technology to cope with it. The technological aids to diagnosis are meant to expand the physician's gaze and to increase precision in testing. However, with an abundance of technologically generated data, the possibility of misinterpretation increases. Even without misinterpretation, multiple plausible interpretations that are consistent with the data are possible, and this can be even more confusing. All this is Wulff's "data pollution."

The use of automated chemical analyzers, for example, has made it easy to get a battery of tests from a single small sample of blood. This, however, encourages the physician to consider a host of data, which at best may be irrelevant to the biomedical problem of the particular patient under consideration, and at worst might be misleading. When a test result is not what is expected or when it is incongruent with other tests or clinical findings, the physician tends to repeat it in an attempt to show that the original unexpected test result was some sort of laboratory error or other aberration. This contributes to the overall glut of data, and forces the physician to choose which data are important according to some set of criteria other than the supposedly objective findings of the laboratory. The problem, however, is that there is no privileged standard by which to make this judgment. One must decide according to some chosen values. Most often what is done is to consider whether the problem test result fits the overall clinical picture as given by the history, physical examination, and other tests. How these different elements are balanced in coming to an overall clinical picture will reflect the many sorts of values we have been considering.

Interpretation of Quantified Data
The levels of various body constituents can be measured to a high degree of accuracy and precision. The reporting of such results by the laboratory can tend to suggest a degree of significance in the measurements that is not borne out clinically. This sort of overinterpretation reflects a system of values in which the assumed objectivity of technologically produced data is favored over the subjectivity of observation, and quantified data are valued more than data specified simply by quality.

Quantified data can be very helpful in clarifying physical findings that are ambiguous. When there is a question of whether a patient is jaundiced, measuring bilirubin can be useful in establishing a diagnosis. However, it is not obvious that knowing that a patient's bilirubin level is 4 milligrams per deciliter is clinically more significant than simply observing that the patient is jaundiced. Being able to say that one's observation has been confirmed quantitatively may be psychologically reassuring, but in many cases where physical findings are obvious, a diagnostic test adds no significant new information. Knowing whether the bilirubin is conjugated or unconjugated, on the other hand, is something that cannot be ascertained by physical examination and may be clinically useful.

Another pitfall in the interpretation of quantified data is related to the problem of distortions due to expectations in the physical examination. The diagnostician may follow the researcher in practicing "data torturing," which can take two forms: opportunistic and Procrustean (Mills, 1993, pp. 1196-1199). In the opportunistic type, the researcher pores over data until some "significant" association between variables is found, and then devises a plausible hypothesis to fit the association. This type may not play a significant role in diagnosis, for diagnosis aims at the practical end of therapy and prognosis and not just at creating facts. In Procrustean data torturing, however, some hypothesis is chosen, and the data are statistically manipulated in such a way as to support the hypothesis. This form may play a role in diagnosis. Tests are chosen so as to provide support for or to rule out diagnostic hypotheses. The diagnostician does not approach the test data blindly, but in light of a particular list of hypotheses being considered. The diagnostician will attempt to see patterns in the data that correspond to the chosen hypotheses.

The Concept of the Normal Range

We have already considered some problems with the idea of normality in medicine when we discussed the concept of disease. The concept of normality is ambiguous; normality might mean what is most common, what is average, or what is ideal (Vácha, 1978, pp. 827-830). Many of the same issues arise when trying to determine the meaning of a normal result of a diagnostic test. Knowing the normal range of a quantitative test is essential for interpretation of the test, yet the very concept of the normal range remains unclear. Ellis Benson (1972, p. 152) has remarked:

> The normal range has had a vague but comforting role in laboratory medicine. It looms on the horizon of our consciousness, perfectly symmetrical like a Mount Fujiyama, somewhat misty in its meaning, yet gratefully revered and acknowledged.

The normal range of any laboratory test is generally based on an assumption that the distribution frequency of test results from a representative sample of healthy people falls into a Gaussian, or "normal" distribution. The normal range is then taken to extend two standard deviations to either side of the mean; this normal range comprises approximately 95 percent of the test results from the population tested, a population assumed to be normal (Wulff, 1976, p. 40).

There are several difficulties with such statistically constructed norms, however. Jiří Vácha (1985, pp. 339-367) has shown how the German constitutional doctrine of the 1920s and 1930s attached a secondary meaning—healthy—to the statistical concept of normality. He rightly points out that this move is an important one for those who would hold a value-free conception of health and disease. After this move is made, health and disease become just a matter of statistics. This will not do, however, as Vácha realizes, for test values that are optimal with respect to health are not always near the mean. The "healthiest" serum cholesterol levels, for example, are significantly below the mean for "healthy" populations.

The assumption in such statistically constructed norms is that the range of measurements falls in a normal distribution. However, distributions thought to be Gaussian might in fact be skewed. If this is so and one does not know it, one's interpretation of laboratory data may be in error.

Another difficulty lies in the selection of the population that is used as the source of the "normal" distribution. Many normal ranges have been obtained epidemiologically from small groups of hospital employees that are assumed to be normal. The normal range of a test may differ in various hospitals; thus, a test value that is normal in one place may be abnormal in another (Feinstein, 1967, p. 333). This might reflect different methods of testing, but it might also reflect the fact that different populations have been used to determine what is normal. One generally cannot know the source of these differences without a great deal of research.

Still another pitfall can arise from a failure to recognize the impact of selection bias in interpreting test results. J. A. Knottnerus and colleagues show how diagnostic habits can serve a self-fulfilling prophecy. Incorrect conclusions can be drawn about relationships between symptoms and diseases because of bias in the selection of comparison groups. Suppose, for example, that there is actually no relation between fatigue and anemia in the general population. However, physicians have learned that fatigue is a symptom of anemia. Therefore, a physician is more likely to obtain a hemoglobin level in a fatigued patient than in a non-fatigued one. Now, the probability of having anemia given that one is fatigued may be no greater than the probability of having anemia in the general population. However, the probability of being fatigued given that one is anemic will appear greater because of the selection bias in favor of fatigued people in the test group. Just such a bias may be what went into the learning in the first place that fatigue is a symptom of anemia. A further conceptual difficulty makes matters even worse. Clinical entities such as fatigue may change in meaning over time, making comparisons of present test results with past data difficult (Knottnerus, Knipschild, and Sturmans, 1989, pp. 67-71).

Finally, it must be remembered that when a continuous variable is being measured, and what is desired is a test result that is either normal or abnormal and not indeterminate, a separating line must be drawn somewhere along the continuum. Diagnostic values will determine where the separation is made, and hence what is considered a "good" or "bad" test result. Where the line that separates normal from abnormal is drawn will also depend on the whether it is more important to avoid false positives or false negatives. This may have to do with the consequences of missing conditions that are considered serious but easily treated, or of labeling someone as having a disease that they do not really have.

Value judgments are thus inherent in the most fundamental aspects of the design and interpretation of laboratory tests. These diagnostic values go beyond the values of the previous three levels and into the immediately moral realm. Diagnostic values go beyond our theoretical musings about facts, values, health, and disease; they have direct impact on the lives of suffering people. The assertion of Georges Canguilhem (1991, p. 131) is apt: "it is life itself and not medical judgment which makes the biological normal a concept of value and not a concept of statistical reality."

Predictive Value of Test Results
Establishing particular diagnoses depends a great deal on some of the statistical models that go into the interpretation of laboratory tests. To understand how this is so, it will be useful to consider briefly the notions of sensitivity, specificity, and predictive value of test results.[4]

To speak of the sensitivity and specificity of a test assumes that there is some standard to decide whether or not a disease is present. The sensitivity and specificity of the test, then, are measures of how well the test does in discriminating presence and absence of disease. We have already considered the philosophical difficulties in establishing such a standard. It may be conceptually naïve to assume such a standard. But suppose we could find such a standard, or suppose that agreement about concepts and values has allowed us to specify such a standard. Even then, additional conceptual commitments and values must come into play in this interpretation.

First, I offer some standard definitions. A *true positive* (TP) is a positive test result in a patient with the disease in question. A *true negative* (TN) is a negative test result in a patient without the disease. A *false positive* (FP) is a positive test result in a patient without the disease. A *false negative* (FN) is a negative test result in a patient with the disease. The *sensitivity* of a test is the probability that a patient suffering from the disease will have a positive test result. The *specificity* is the probability that a patient not suffering from the disease will not have a positive test result. So,

$$\text{Sensitivity} = \frac{TP}{TP + FN}$$

and

$$\text{Specificity} = \frac{TN}{FP + TN}$$

What this shows is that a very sensitive test will be good at identifying those people who actually have a disease. A very specific test will be very good at identifying those people who do not really have the disease.

The predictive value of a positive test result is the probability that a positive result is truly positive. That is,

$$\text{Predictive value of positive test} = \frac{TP}{TP + FP}$$

Likewise, the predictive value of a negative test result is the probability that a negative test result is truly negative. That is,

$$\text{Predictive value of negative test} = \frac{TN}{TN + FN}$$

The importance of the predictive value of a positive result is evident, for a positive result will most often lead to a diagnosis, which serves as a justification for a particular treatment. It also leads to the labeling of a person as having a particular disease, which may have important social ramifications.

It can be further shown, by simple conversions of these terms into Bayesian terms and substitution into Bayes's Theorem, that the predictive value of a positive test result (PV(+)) depends on the prevalence of the disease in the population (Statland, et al., 1979, p. 529). That is,

$$PV(+) = \frac{Prev \times Sens}{(Prev \times Sens) + [(1 - Prev) \times (1 - Spec)]}$$

where "Prev" is the prevalence of the disease in the population, and "Sens" and "Spec" are the sensitivity and specificity of the test, respectively. What this means is that the lower the incidence of a disease, the lower the predictive value of a positive test result will be. Take the example of a test that is 95 percent specific and 95 percent sensitive for some disease. If the prevalence of the disease is 50 percent, the predictive value of a positive test result is 95 percent. If the prevalence were only 5 percent, however, the predictive value would be only 50 percent. With a prevalence of 1 percent, a test with the same sensitivity and specificity would have a positive predictive value of only 16.1 percent. Such a test might be a good diagnostic test in a unit that specializes in that particular disease, but it would be a poor screening test in a population with a very low incidence of the disease (Statland, et al., 1979, p. 530).

Most laboratory tests measure what they measure on a continuous scale. A practical difficulty that this raises is deciding on the "cutoff value" that separates a positive and a negative test result. The sensitivity and specificity depend on the cutoff value selected, and so the predictive value of the test depends on the cutoff value. When a cutoff value is changed, either sensitivity is sacrificed for the sake of specificity, or specificity is sacrificed for the sake of sensitivity. One must decide which is more important. This will depend on the various diagnostic values we have already discussed. Hence, the choice of a cutoff value is necessarily a normative activity.

Experience with the complexities of clinical medicine may actually affect how a physician reasons about these matters. In one study, the following question was asked to medical students and residents:

If a test to detect a disease that occurs in one of one thousand people is falsely positive in 5 percent of unaffected people, what is the chance that a person found to have a positive result actually has the disease?

The correct answer, calculated according to the standard equations given above is 2 percent. Eighty-six percent of first year medical students answered the question correctly. Third year students did not do as well, interns did worse yet, and residents gave the poorest performance, with only 48 percent answering correctly. The same problem was framed in terms of a test for a problem with an automobile:

> If automobile inspection to detect a serious fault that occurs in one of one thousand cars also fails 5 percent of automobiles without the fault, what is the chance that a car that fails inspection actually has the serious fault?

With this problem, the residents did significantly better, with 59 percent answering correctly (Holtzman, 1994, pp. 241-243). It appears that clinical experience colored the residents' thinking about purely logical questions so that they did not take advantage of the objectivity that the test was supposed to provide.

As we have already seen, the diagnostic process is guided by the aims for making a diagnosis. Lars Elffors (1988, pp. 146-148) suggests some possible aims: (1) as a guideline for choice of therapy, (2) as a tool for making prognoses, (3) as a primarily scientific act, and (4) as a formal medico-legal act.

With respect to the first two reasons for diagnosis, the preceding mathematical analysis has great relevance. Suppose a diagnostician is faced with a choice between two laboratory tests for some disease: test A has a high sensitivity and low specificity, and test B has a high specificity and low sensitivity. The choice of the proper test to use depends upon the disease in question, which in turn depends on conceptual and normative commitments. Test A, highly sensitive and less specific, would be desirable for detecting diseases that are serious and treatable. The consequences of failing to diagnose such a condition are serious. The false positives that will result from the test's low specificity are tolerated because it is most important that the treatable condition not be missed. Test B, on the other hand, would be desirable in the diagnosis of a condition that is serious but not treatable. Here a false positive would be very undesirable because it would cause the patient great distress to wrongly believe that he or she were suffering from a serious and incurable disease. Likewise, if a treatment for the disease were particularly risky, one might prefer test B, for a test with high specificity would rule out treating a person who did not really have the disease.

The value judgments in such diagnostic decisions are not hard to see. One must consider not only what sorts of physical harms one is willing to tolerate, but also psychological and economic harms as well. Some such values may be more or less universally held; others may be idiosyncratic. There may be differences between patient and physician in this regard. This points to a need for discussion between physician and patient before diagnostic interventions as well as treatments. Judgments must be made about when the information that might be gained from a diagnostic process is worth the risk involved (McCullough and Christianson, 1981, p. 141).

When it comes to making diagnoses in clinical research, it will often be of primary importance to use a highly sensitive test. In such research environments, especially when a study population is relatively homogeneous, the prevalence of the disease under consideration will be relatively high, thus giving positive test results a high positive predictive value.

With respect to medico-legal diagnoses such as random drug screening, the incidence of the condition in the population is likely to be low. In this situation, using a very sensitive test would produce a large number of false positives and would wrongly label many people as engaging in an illegal activity. This could have serious social implications for them. Again, value judgments must be made about the tradeoffs in social and legal consequences of missing diagnoses, and falsely attributing diseases to people who do not have them. Engelhardt (1980, p. 45) sums this up:

> Establishing any particular level of substantiating data for a diagnosis as a criterion for making a particular diagnosis reflects a judgment about how one should prudently balance the goods versus the harms of false positive versus false negative diagnosis.

V. CONCLUSIONS

We have seen that values are inherent in the three major elements of diagnosis: the history, the physical examination, and the diagnostic tests. Interpretation of facts in all these areas will be informed by the values of the interpreter, that is, the physician. The patient, however, is also an interpreter in the diagnostic process. The patient's narration of his or her illness will be an interpreted narration, interpreted in light of a set of values that will most likely differ in at least some important aspects from the physician's set of values. The patient also steers the diagnostic process by consenting to or refusing certain diagnostic tests. This consent is ethically mandatory because of the risks that are inherent in certain diagnostic tests. Thus, the empirical facts that are gleaned in the diagnostic process are limited by the extent of the patient's consent to be tested. The limitations that patients place on the process reflect their values in many realms. Thus, unlike foundational and nosological values, where the value input is primarily from the scientific and professional arenas, diagnostic values, like conceptual values about disease, importantly involve the values of patients as well.

What remains for us to consider is the way in which physicians put together what they have learned in the history, physical examination and ancillary tests to make a diagnosis. This process is the subject of the next chapter. We will see that the theory behind this process is complex and controversial. However, it will be evident that diagnostic values necessarily play an important role.

CHAPTER 7

THE PROCESS OF DIAGNOSIS

I. INTRODUCTION

In the last chapter we considered the elements of diagnosis: history, physical examination, and various types of diagnostic tests. Out of these elements come the data used to construct diagnoses. In this chapter I will examine how these data are used in the construction project.

First, I will consider some definitions of diagnosis. I do not think it is possible to give a precise definition of diagnosis because the elements of diagnosis and the data derived from them are not precisely definable. However, it is worth discussing definitions of diagnosis to emphasize the overall thesis of this book: that diagnosis cannot be understood in terms of value-free scientific facts.

Next, I will discuss the logic of diagnosis. The way physicians actually carry out the diagnostic process is not well understood. We will see that the diagnostic process is too complex to be described in terms of any one type of logic.

I will then consider intuition in the diagnostic process. The sort of intuition that can play a valid role in diagnosis cannot be any sort of mystical grasp of reality. Intuition in the sense of an "intellectual shortcut" may play an important role when a diagnosis must be made urgently. This sort of intuition, in my view, is tied to expertise within a particular worldview, and so I will examine the role of expertise in the diagnostic process.

Finally, I will look at the use of computers in diagnosis. The problem with computer diagnosis at present does not stem from the computer's inability to handle the value-laden nature of diagnosis. It should be possible to devise a computer program that is able to deal with values as well as facts. The problem lies more in our failure to adequately describe the diagnostic process in terms of a logic that is precise enough for computers to handle.

In all of this I will show how the diagnostic values we identified in the last chapter necessarily come into play in the diagnostic process. In the end, the act of diagnosis creates a new fact about a patient. This fact, however, is embedded with the foundational values of scientific theories, the conceptual values of the concept of disease, the nosological values of our classifications of disease, and the diagnostic values of the process of diagnosis.

II. THE DEFINITION OF DIAGNOSIS

Stedman's Medical Dictionary (1995) defines diagnosis as "the determination of the nature of a disease." This dictionary definition takes diagnosis to be an action, not a generic noun standing for the name of a particular disease. Common usage, however, often favors the latter, as in saying that someone has a diagnosis of leukemia. I think it unwise to try to eliminate this ambiguity by introducing a new word for the diagnostic process, such as "diagnostics," which itself introduces an ambiguity since it is also used for the study of the diagnostic process. Context usually makes the meaning clear enough. When clarification is needed, I will refer to the action of diagnosis as the "diagnostic process."

The dictionary definition has more serious problems, however. We have already discussed at length the ambiguities present in "the nature of disease." Some diagnoses do not describe the nature of disease at all, but simply name symptoms or groups of symptoms.

This, along with some reflection on the goals of medicine, has prompted some revised definitions of diagnosis. Several of these have rightly pointed out that diagnosis is done for the sake of other goals. Henrik Wulff (1976, p. 78) says:

First of all it should be remembered that the diagnosis is not an end in itself; it is only a mental resting-place for prognostic considerations and therapeutic decisions. . .

H. Tristram Engelhardt, Jr. expands this idea that diagnosis is not an end in itself. He sees diagnosis as involving an inseparable mixture of descriptions, explanations, predictions, and evaluations. In diagnosis, a patient's condition is described, usually in terms of established nosographic terms (signs and symptoms) and explained in terms of some disease entity. However, the particular account chosen depends on the likely consequences of the therapeutic interventions that it would justify, and on an assessment of projected outcomes given some particular treatment or no treatment at all (Engelhardt, 1980, p. 45). It is important to note that moral values, in addition to the scientific, professional, and aesthetic values inherent in the construction of disease and disease classification, are made overt in connecting diagnosis with the broader goals of medicine. Diagnostic tests, as we have seen, involve putting people at risk for different sorts of harm and thus are in the moral realm. Linking diagnosis with prognosis and projected treatments brings diagnosis deeper into the moral realm.

Caroline Whitbeck likewise argues that diagnosis has no goals of its own that are separable from the general goals of clinical medicine. Diagnosis, like all of clinical medicine, aims at the prevention and treatment of disease, which will result in the best outcome for the patient. Thus, when it is best for the patient, it is sometimes preferable to settle for diagnostic categories that are not the names of disease entities, but are merely names of symptoms. To arrive at such a "nonspecific" diagnosis as "flu" may be good diagnostic practice when it is not "clinically worthwhile" to further classify the condition (Whitbeck, 1981b, pp. 319-322). Several sorts of value judgments will be involved in determining what is clinically worthwhile. Some will reflect more or less universally accepted values, such as those concerning pain and death, and some will reflect the personal values of the individuals involved.

Whitbeck (1981b, p. 324) suggests this definition of diagnosis:

> Diagnosis is the process of inquiry aimed at discovering *the causes and mechanisms* of a patient's disease insofar as this information is needed to inform treatment and management decisions to achieve the best medical outcome for the patient, and to prevent disease in others.

This definition has the merit of putting diagnosis in its proper context of clinical medicine, as opposed to the discovery process that is more proper to basic science research. In addition, it clarifies to some extent the dictionary reference to the "nature of disease" in its specification of causes and mechanisms as the goal of discovery. Whitbeck's definition makes explicit the value content of diagnosis in its reference to the "best" medical outcome. However, the definition fails to make explicit exactly *how* evaluative elements come into play in diagnosis. For instance, what is the "best medical outcome"? Does this reflect the patient's values, or is it the physician's judgment about what is "medically" best. The two might not be the same.

To that end, the analysis of Kazem Sadegh-Zadeh is helpful. Sadegh-Zadeh sees diagnosis as a *categorical statement* of particular structure about an individual. An example is "David has bronchitis." This simple statement, however, is in fact a complex opinion that involves a number of factors. Sadegh-Zadeh understands a medical diagnosis in terms of a seven-place predicate:

$MedDiag(\alpha,t,x,y,W,M,D)$.

This is read as "α is a medical diagnosis at t about x relative to y, W, M, and D," where α is the diagnosis, t is a point in time, x is the patient, y is the physician, W is a conceptual system, M is a diagnostic method, and D is a set of data (Sadegh-Zadeh, 1981, p. 189-191).

Sadegh-Zadeh adds some other conditions in an attempt to make the definition of diagnosis more precise: that the diagnosis cannot entail a contradiction, for example. These details need not concern us here, however. What I want to emphasize is how conceptual and value commitments are embedded in the definition of diagnosis as stated.

The diagnosis, α, is a function of six things (Sadegh-Zadeh, 1981, pp. 187-188). First of all, the diagnosis is obviously related to a patient, x. A diagnosis is always a diagnosis with respect to a particular individual. Second, the diagnosis involves a physician, y, who carries out the diagnostic process. Diagnostic competence on the part of the physician is expected if a particular judgment is to be considered a diagnosis. I will consider this in more detail later. Third, the diagnosis is always a diagnosis at a particular time. A statement that constitutes a diagnosis for x today (e.g., x is having an acute asthma attack) may not constitute a diagnosis for x tomorrow. This reflects a particular ontological conception of disease in which time is as important as space. Fourth, a statement's being a diagnosis depends upon a particular conceptual system. This conceptual system will include all the matters we considered in chapters 2 through 5, including worldview about scientific theories, the concept of disease (which Sadegh-Zadeh explicitly recognizes), and particular nosologies. Along with all these elements go the values, which are inherent in them. Fifth, the diagnosis is dependent upon the data from which it is constructed. A particular set of data must be chosen out of a massive number of potential data. Different sets of data are very likely to give rise to different diagnoses. The act of choosing particular data as relevant introduces all the conceptual and value commitments we discussed in the last chapter. Finally, a diagnosis depends on the method used to generate it. Some methods are better than others are. A statement arrived at by tossing a coin or practicing sorcery will rightly not be considered a diagnosis (Sadegh-Zadeh, 1981, pp. 187-188).

This analysis of diagnosis does not make explicit all the values that go into the construction of a diagnosis. However, it does make explicit many of the conceptual issues that come into play in the diagnostic process but remain hidden in the dictionary definition of diagnosis and in Whitbeck's definition.

Value judgments, however, also have a prominent role in diagnosis. A medical diagnosis is a description of some properties of a person, but the diagnosis will include only those properties that are interesting from a medical point of view. This means omitting other properties. As Geo Săvulescu (1976, pp. 328-329) says:

> In most cases, the diagnosis is not complete, but partial. When we diagnose a certain patient as suffering from ankylosing spondilities [*sic*], we cannot consider it a complete diagnosis. To have it complete we should make clear if patient x or y suffers from other illness or not, we should note the clinical form of the spondilities [*sic*] ... the patient's age and sex as well as other individual determinations of the illness which can be of interest.

The necessity of invoking what is "of interest" in formulating a diagnosis reveals the implicit value judgments that must be made in the diagnostic process. What is of interest in the clinical encounter will be relative to the values of both patient and physician. Some of these values may be individual and others may be more or less universal, such as those involved in the notions of function as discussed in chapter 3.

III. DIAGNOSTIC METHOD

Many attempts to describe an exact diagnostic method are motivated by the desire to make diagnosis more objective, more precise and more accurate. Sadegh-Zadeh sets out conditions for an exact diagnostic method (Sadegh-Zadeh, 1981, pp. 192-194). An exact diagnostic method applied by different diagnosticians to a given patient should yield the same diagnosis. This is so, however, only if the data sets and conceptual commitments of both diagnosticians are the same and both diagnosticians work simultaneously. I would add that the value commitments of both must also be the same. However, as Sadegh-Zadeh (1981, p. 191) points out, the ideal of an exact diagnostic method cannot be developed because there is no way to eliminate from the diagnostic process the patient and the physician, both of whom "possess individually different sagacities, sensitivities and inexactitudes peculiar to human beings." Many of these sensitivities and inexactitudes, in both patient and physician, are related to individual values.

The Problem of Vagueness

At least part of the difficulty we find in trying to describe a diagnostic method precisely lies in the number of vague terms that are used in everyday clinical practice:

> many, most, few, often, seldom, very often, very seldom, frequently, occasionally, sometimes, almost never, almost always, rather, common, most common, typically, acute, chronic, sub-acute, sub-chronic, sub-clinical, rapidly, slowly, suddenly, may, can, possibly, perhaps, small, great, mild, moderate, severe, cyanosis, red, yellow, pain, headache, malaise, tender, flabby, hepatomegalia, polyuria, oliguria, etc. (Sadegh-Zadeh, 1982, p. 110).

The philosophical issues surrounding the concept of vagueness are complex and controversial.[1] Note, however, that such words as "flabby" have borderline cases. How soft does something have to be before it becomes flabby? Some tissues are clearly flabby and some are clearly not flabby, but it is never certain when the borderline of flabbiness has been crossed. The same is true for observations of color. We may clearly recognize something that is red, and if the color changes by just a small bit, we will still recognize it as red. However, a series of such small changes will eventually lead us from red to orange. The line between red and orange is similarly vague. Crispin Wright has used the term "tolerance" to describe such predicates. A Predicate F is tolerant with respect to ϕ if there is "some positive degree of change in respect of ϕ insufficient ever to affect the justice with which F applies to a particular case" (Wright, 1996, pp. 156-157).

Russell (1996, p. 64) suggests that many of our observational predicates are tolerant in Wright's sense because of our physiological constitution. For example, "red" is a term applied to a sensation. When we have two indistinguishable sensations and we call one "red," we must also call the other "red." It is unclear, however, whether the vagueness resides in the limitations of our unaided observational powers, or in the stimuli themselves.

Different theories of vagueness answer this question in different ways. I will only discuss one set of opposing theories here; this is all we need to understand what is at stake in the present context of diagnosis. The simplest approach to vagueness is the *epistemic* view, which comes to us from the Stoics. According to this view, vague predicates do have well defined extensions; we simply lack the knowledge of how to draw the boundaries precisely (Cicero, *Academica* 2.92-94, trans. Rackham, H.). James Cargile (1996, pp. 96-98) calls this view *realist*. This is opposed to a *nominalist* view, which holds that we simply draw arbitrary boundaries to separate, say, red from non-red.

The naïve realist about disease and diagnosis would be likely to hold the realist view of vagueness. Using laboratory tests in an attempt to eliminate the epistemological ambiguities of physical examinations seems to presuppose such a realist view of vagueness. On this view, physical characteristics are not vague in themselves. Vagueness comes from our inability to apprehend them in their proper detail.

We need not settle this deep problem here. Either way, values must be brought into our analysis. If the nominalist view is correct, we will incorporate our conceptual, nosological, and diagnostic values in our decisions about where to draw our arbitrary boundaries. If the realist view is correct, our epistemic uncertainty will still force us to draw boundaries as we struggle to classify vague terms; we will still need to incorporate values. As we saw in chapter 6, where we draw these boundaries will depend on whether we are more concerned about false positives or false negatives.

Attempts to make the language of diagnosis more precise might be desirable for increasing the precision of the diagnostic process, but vague terms constitute many of the building blocks of the most basic concepts in the worldviews of diagnosticians. It would be impossible to eliminate all vagueness from the diagnostic process without fundamentally altering some of the basic concepts on which the diagnosis depends. This is true especially of the patient's experience of illness.

The Logic of Diagnosis

Diagnosis is an extremely complex problem-solving exercise, in part because of the sheer number of biomedical problems that might possibly be consistent with the signs and symptoms of a patient, and in part because of the complexity of diagnostic tests themselves. No consensus has been reached concerning the logic of clinical diagnosis. However, it is generally agreed that a more rational and systematic approach to diagnosis is desirable.

Elliott Sober (1979, pp. 29-44) considers four reasons why diagnosis has been considered an art, as opposed to a science, and attempts to refute all of them. This endeavor is a useful attempt to show that diagnosis can be a rational process. However, we must keep in mind that science itself involves an element of what we would properly consider art. There are better and worse ways, from an aesthetic point of view, to reach the conclusions at which the practice of scientific medicine aims. Thus, we cannot maintain a strict dichotomy between science and art.

First, Sober refutes the claim that because each individual is unique, no formal logic can capture the diagnostic procedure for that individual. Such a view of maximal uniqueness would make any sort of inductive generalizations about an individual impossible. But, says Sober, we do make generalizations and apply them to individuals, and these generalizations often turn out to be useful. Sober is correct that this claim does not exclude diagnosis from the realm of science. Just because we make useful generalizations, however, does not make diagnosis pure science as opposed to pure art. We make many useful generalizations about aesthetic values, but this is not to say that aesthetic judgments are a science.

The second reason concerns matters of ethics, which Sober sees as outside the realm of science. But the ethical dimension does not remove diagnosis from the realm of science, for all of science has a value dimension. Some of these values, as we have seen, are moral values. The ethical aspect only points out the need for moral considerations to be part of one's "rational calculations," as Sober puts it, about the extent people will be subjected to diagnostic testing.

The third factor is the dichotomy between rationality and the emotions. But this may be a false dichotomy. Some might deny the role of the emotions in the ethical aspects of diagnosis. But suppose that an ethical approach to diagnosis does demand empathy and sensitivity for the patient in the diagnostic process. This seems to be a reasonable supposition. However, it need not make the diagnostic process unscientific or irrational; it merely points out the importance of the emotions within a scientific process.

The fourth factor is the quantitative-qualitative dichotomy. However, there is no reason to believe that quantitative information is necessarily more precise and "scientific" than qualitative information. Quantitation is indispensable in many circumstances. However, quantitation can also provide needless or even misleading precision, as when small deviations from a normal range are given importance they do not deserve.

These issues give some indication of the difficulty of specifying a formal logic of diagnosis. It may be that diagnosis cannot be formalized according to one system, and that it involves several different ways of thinking.

Several logics of diagnosis have in fact been advocated. Some of these claim to be merely descriptive of the way physicians actually carry out the diagnostic process. Others claim to be prescriptive, that is, they make normative claims about how the diagnostic process *ought* to be carried out.

The most prominent of the prescriptive methods are the Bayesian and branching logic approaches. I do not believe that either of these is adequate as a complete formal model of clinical diagnosis. Each, however, can help the cause of rational diagnosis when used in particular facets of the diagnostic process.

Bayesian Models

Bayesian models give a formal means of revising opinions about the probability of the presence of a disease in light of new information. Bayes's Theorem, in its simplest form relating to diagnosis, can be stated as follows:

$$p(D \mid E) = \frac{p(E \mid D) \times p(D)}{p(E)}$$

where $p(D|E)$ is the posterior probability of the presence of disease D given evidence E, $p(E|D)$ is the probability of obtaining E given the presence of D, $p(D)$ is the probability of the occurrence of D, and $p(E)$ is the probability of obtaining E in the population (Cohen, 1980, p. 47).[2] What Bayes's Theorem tells us is that, given a new bit of evidence, the probability of a patient's having a particular disease is directly proportional to the probability of the disease in the population and to the probability of obtaining the evidence given that the disease is actually present, and inversely proportional to the probability of finding the evidence in the population as a whole.

In a Bayesian approach, the estimate of the probability of a particular diagnosis in light of particular data depends directly on the probability of finding the disease in the population. Access to accurate information about this latter probability, however, is rarely available. As Henrik Wulff points out, this estimation will depend on the experience of the particular clinician, as well as on the setting in which the clinician practices. In a university hospital setting, where the incidence of some rare disease D is relatively high, a bit of clinical evidence for D may be a fairly good predictor of D. However, if the clinician were to move out of the university hospital and into a general practice setting, the prevalence of that rare disease would be much lower. If the clinician were to use the hospital-based $p(D)$ to calculate the probability of a general practice patient's having D given the same evidence as before, the resulting probability would be too high (Wulff, 1976, pp. 83-84). In practice, most diagnoses are constructed mainly from the results of diagnostic tests. There is no independent way of establishing the prior probability of the disease in the population.

Bayes's Theorem can be given in more complex and sequential form, which allows one to update the probability of a disease given multiple bits of new information such as laboratory values and further physical findings.[3] Given the usual sequential nature of discovery of evidence in the diagnostic process, this form would apparently be much more useful. However, Bayes's Theorem assumes that each bit of evidence is conditionally independent. This assumption is generally violated in medicine, as Kenneth Schaffner (1981, p 169) shows. Individuals with congestive heart failure will manifest edema and an enlarged liver, but these manifestations are not conditionally independent.

The degree of independence of manifestations will vary with the type of disease. The syndrome provides a good example. The manifestations of a syndrome that is poorly understood from a physiological point of view may well be independent. However, as the syndrome acquires some explanation in terms of physiology, the list of the syndrome's manifestations may change, and those manifestations that remain in the syndrome description may then be seen as together dependent on some common underlying phenomenon. Thus, as a syndrome becomes better understood, the assumptions of Bayesian methods may become less justified.[4]

Schaffner (1981, pp. 169-170) also raises another difficulty in the Bayesian approach. The set of diseases in question often cannot be partitioned in an exhaustive and mutually exclusive manner. This problem is related to the issues of the metaphysics of disease that we have already discussed at length. A set of data may point to one disease or to the presence of several concurrent diseases. It will prove exceedingly difficult to establish probabilities for all the possible combinations of diseases given a complex set of data. As we have observed, it is possible to get from a particular patient a large set of descriptive data. How those data are partitioned into facts about the patient will depend on all the values we considered in our discussion on disease classification. However, sharp boundaries cannot always be drawn between disease categories. A Bayesian analysis will not be helpful in delineating sharp boundaries between entities that have no sharp boundaries.

Moral values are implicit in Bayesian methods of diagnosis. An evaluation of a diagnosis in terms of probability must accept some rate of error. Some body of evidence E_1 might support a 95 percent probability of diagnosis D. However, further evidence E_2 can always be found, and $p(D|E_1\&E_2)$ might be much less than 95 percent (Cohen, 1980, pp.53-55). It is always possible to obtain more evidence, and one cannot be sure that some future evidence will not significantly weigh against a prior probability. One must decide what degree of probability warrants a diagnosis that may have serious implications for the patient. Thus, moral values, as well as scientific and epistemic values, come into play in such decisions.

Branching Logic Models

Other authors argue that alternative models, which make use of flow charts and branching logic, better exemplify actual clinical reasoning. Alvan Feinstein (1973b, p. 267) finds three distinct components in the selection process of choosing an appropriate final diagnosis: a set of candidates, the mechanism for "nominating" or excluding candidates in the set, and a mechanism for excluding or "ruling out" all but the ultimate choice. He sees diagnostic reasoning as the "process of converting observed evidence into the names of diseases." The format of this reasoning is, on his account, a "branching series of logical decisions, each of which produces intermediate conclusions during the progressive transformation of input to output." (Feinstein, 1973a, p. 212). Diagnostic reasoning consists in "sequential stations" which can be graphically demonstrated in a flow chart. The physician begins by noting a manifestation in a patient for which a diagnostic explanation is sought (e.g., difficulty in swallowing). The manifestation is then referred to a domain (e.g., upper gastro-intestinal tract) and to a focus in that domain (e.g., pharynx or esophagus). The next decision is to choose a disorder at that focus. Other ancillary disorders may be invoked in parallel or sequential stages of the reasoning (Feinstein, 1973a, p. 224).

Benjamin Kleinmuntz (1968, p. 178) represents diagnostic search strategies as "binary tree structures" consisting of a number of test nodes which seek to elicit data by questions that move from general to specific. Such flowchart programs can be used to guide the diagnostic process in a step-by-step fashion. In their simplest form, they represent what Szolovits and Pauker call "categorical" reasoning (as opposed to probabilistic reasoning). Categorical decisions depend on relatively few facts and have unambiguous results. The problem is that physicians most often work with categorical decisions, even though many decisions they routinely make (e.g., whether to perform a bone marrow biopsy) are too complex to be handled in this fashion (Szolovits and Pauker, 1978, pp. 117-119).

Branching logic methods need not be purely categorical, however. A branching tree structure can have "chance nodes," which incorporate Bayesian probabilities at certain points. This approach improves the ability of the system to handle complex bits of information, but also introduces all the problems of Bayesian analysis.

An additional general problem with branching logic models is that the sequence of decisions in this method is extremely rigid, making it difficult to reverse a path of reasoning in light of new information (Schaffner, 1981, pp. 170-173). This is a problem in diagnosis because data relevant to decisions at early branch points may become available only later in the diagnostic process.

Cohen contends that the branching logic mode of evaluation is essentially what he calls "inductive, or Baconian, probability." It is not clear that branching logic really is inductive reasoning. However, Cohen's point is that the reliability of a generalization is best evaluated by the variety of potentially relevant circumstances that fail to falsify it. Ideally, we ought to be able to rank-order any set of diagnoses within a particular field of medicine according to the weight of evidence in favor of them. So, an idealized reconstruction of diagnostic reasoning ought to eliminate more hypotheses in the beginning stages of inquiry than in the later ones. Bayesian reasoning might be helpful in establishing these diagnostic weights (Cohen, 1980, pp. 56-60).

Probabilistic and categorical reasoning both play an important role in the diagnostic process. Furthermore, the processes of diagnosis and treatment are often concurrent, with information garnered from each simultaneously affecting thought about the other. Feinstein (1974, p. 30) concludes:

> Diagnostic algorithms include the analysis of paraclinical data, the analysis of clinical data, and the plans for a diagnostic workup. Although purely diagnostic algorithms may be followed by separate algorithms for prognostic and therapeutic reasoning, these procedures may often occur in an intermingled sequence in clinical practice. The diagnostic process may be interrupted by treatment that acts as a diagnostic test or that eliminates the need for further diagnosis. The traditional diagnostic succession of history, physical examination, and paraclinical tests may also be performed in a sequence different from conventional pedagogic instructions.
>
> The algorithmic portrayal of these processes is crucial for determining the intellectual "economy" with which they are performed, for improving the way in which they are taught, and for assembling satisfactory data to evaluate their costs, risks, and benefits to patients. The construction of justified clinical algorithms requires intimate familiarity with clinical activities and offers a major new scientific challenge in basic clinical research.

Feinstein rightly points out the intimate connection of diagnosis with prognosis and treatment, as well as the complexity of diagnostic reasoning. Furthermore, he points to the need for diagnostic algorithms to take into account necessary value judgments about costs, risks, and benefits to patients in the diagnostic process. Thus, in proposing the construction of a normative account of the diagnostic process, Feinstein points to the need for familiarity with different sorts of clinical activities that are value-laden.

I next want to look at some of the studies that have provided descriptive reconstructions of the diagnostic process. In so doing, we will get some insight into the ways in which experienced diagnosticians actually reason.

Hypothetico-Deductive Reasoning
The major study of diagnostic reasoning conducted by Elstein, Shulman, and Sprafka (1978) found that physicians' diagnostic activities can be

summarized by a four-stage model: cue acquisition, hypothesis generation, cue interpretation, and hypothesis evaluation.[5] Cue acquisition refers to the data gathering process. In diagnosis, unlike other problem-solving exercises, all the information needed to solve the problem, that is, to make a diagnosis, is not available to the physician at the beginning of the diagnostic process. Because the prompt generation of a reliable diagnosis is desired, physicians generate diagnostic hypotheses very early on in the diagnostic process, usually as soon as the history-taking process begins, and sometimes at first sight of the patient. Early hypothesis generation permits the use of the hypothetico-deductive method in which cues are interpreted as tending to support or disconfirm a hypothesis, or as noncontributory. As the information gathering continues, the tentative hypotheses are ruled out until only one most likely candidate remains.

The use of a hypothetico-deductive method by physicians in the diagnostic process is akin to Karl Popper's logic of scientific discovery. Popper avoids the problems of inductive logic by claiming that the logic of scientific discovery is deductive. For Popper, predictions that are easily testable are deduced from a theory. If the predictions are verified, then they support the theory, although only provisionally. If the predictions are falsified, however, they definitively falsify the theory from which they were produced (Popper, 1992, pp. 32-33). Inductive logic does not come into play in this scheme.

Hypothetico-deductive logic in diagnosis depends on the conceptual commitments that have gone into the construction of the nosologies that provide the basis for generating hypotheses. These commitments include foundational, conceptual, and nosological values. The hypotheses that are generated will depend upon the learning of the diagnostician and the organization of the diagnostician's medical knowledge. This medical knowledge may be ordered around the central value of helping the suffering patient, or around other purposes. Thus, the diagnostic process only rarely attempts to specify some novel condition; constructing new diseases is more the stuff of clinical research. Routine diagnostic reasoning works with already accepted concepts. However, there are cases in which the diagnostic process may arrive at some novel conclusion, when this seems best for offering treatment or prognosis to the patient.

Graham Bradley reviews several of the studies that point to the use of hypothetico-deductive methods by physicians in diagnosis. The number of hypotheses generated in the diagnostic process can vary, but generally ranges between four and seven. With such a small number of potential diagnoses, it is important that early hypotheses be of a quite general nature so they do not rule out potentially relevant possibilities. Several psychological factors come into play and influence the interpretation of data. In the generation of hypotheses, information gathered early in the diagnostic process is often afforded more weight than later information, and positive findings are often given more importance than negative findings (Bradley, 1993, pp. 54-55). Such weighing of information may bias the logical conclusions that may be drawn from the evidence.

A study of the diagnostic reasoning of neurologists showed that these physicians formulate hypotheses very early in the diagnostic encounter with the patient. The hypotheses are reported to "pop into the clinician's awareness." The neurologist then asks questions that are aimed at achieving specific information. At some point, the diagnostician decides that the inquiry is no longer worth pursuing (Barrows and Bennett, 1972, pp. 273-277). This illustrates a diagnostic value in the diagnostic process. A judgment must be made about whether the epistemic value of potential information justifies the added effort, cost and potential health risks of further diagnostic inquiry. Because the amount of information that could potentially be sought in any given case is virtually limitless, this sort of value judgment plays a role in every diagnostic encounter.

Difficulties in Formalizing the Diagnostic Process

Another study affirms the sequential nature of clinical reasoning and shows that physicians do not actually follow the normative rigor that is called for by Bayesian and branching logic models. A conclusion of this study is worth quoting:

> The physicians we studied did not generate global outlines of their decisions; rather, they incrementally structured a sequence of decisions, which they made using limited information about the alternative choices; they did not use actual numbers to represent their quantitative reasoning; they worried little about the precise values of the outcomes; they failed to consider treating without further testing; and they failed to recognize outcomes of potentially equivalent value. The picture that emerges is one in which decisions are handled consecutively, a pattern that suggests that the use of heuristics (short

cuts, or rules of thumb) dominates clinical problem solving (Moskowitz, Kuipers, and Kassirer, 1988, p. 442).

This study suggests that even though physicians say that they seek more precision in diagnosis, and often obtain quantitative measurements in the name of enhancing objectivity, they do not actually use the quantitative aspects of the results; rather, they convert the quantitative results into qualitative ones.

Jan Doroszewski offers a good reason why quality may be more important than quantity at the stage of hypothesis formation. At this stage of the diagnostic process, it is much more important to identify possible pathological states that are clinically important than to know their probability. All diagnostic hypotheses considered important should be investigated, regardless of their probability (Doroszewski, 1980, p. 180). Suppose some particular diagnosis is highly improbable. If the physician does not initially consider this improbable diagnosis, and the improbable turns out to be the case, the diagnosis will be missed. The consequences for the patient could be devastating. Again, diagnostic values come into play in the assessment of the importance of clinical conditions. It is important to note that particular biomedical problems may differ in importance to physician and patient. The physician will make estimates of importance according to value judgments about severity of the considered diseases and the possible consequences of "missing a diagnosis." A patient may have a different set of values. The patient might be willing to accept the possibility of living with a particular condition rather than going through some diagnostic procedure to rule it out.

Edmund Pellegrino (1979, pp. 182-183) notes that several of the realities of clinical medicine can upset the orderly process of diagnostic reasoning:

the urgency of the patient's conditions, the absence of a diagnostic set or algorithm which fits the presenting signs and symptoms closely enough, lack of specific tests to discriminate among probable diagnoses, incomplete or inconclusive probability estimates, limitations on data selection related to cost or geographic locale, unreliability of the patient as historian, and the dilemma of weighting the probability of a serious vs. nonserious, and treatable vs. nontreatable disorder, when none of these possibilities is definitely excludable.

While Pellegrino is in favor of making the diagnostic process more logically rigorous, he does not believe that all clinical decisions can be reduced to algorithm. The diagnostic question, "What can be wrong?", seems to require different kinds of reasoning. Pellegrino argues that in classifying signs and symptoms into classificatory patterns, mathematical and statistical types of reasoning are most important. However, in his view, the process of differential diagnosis is significantly different, more akin to arguing a case in court than to proving a scientific hypothesis. The purpose of differential diagnosis is to "make one diagnosis sufficiently more cogent than the others so that it becomes a defensible basis for decisive action." Taking this adversarial approach to differential diagnosis may be especially important because of the psychological tendency to want to prove rather than disprove one hypothesis (Pellegrino, 1979, pp. 176-177).

None of this is to say that value judgments cannot be rationally analyzed. Ledley and Lusted (1959, pp. 15-17) suggest that decision analysis according to "game theory" can be used to quantify values in the diagnostic process. Such approaches may hold promise in the area of computer diagnosis, for they would provide a way of overtly incorporating value judgments in the diagnostic process into a quantified decision-making model. It is still an open question, however, whether such schemes would be able to incorporate the full range of values that people bring to the diagnostic encounter. The non-utilitarian ethical theorist will object that values cannot be fully quantified. Even if they could, it is not such a quantification that determines what ought to be done.

IV. INTUITION AND EXPERTISE

Intuition

Graham Bradley (1993, pp. 66-69) plausibly argues that physicians do not have time in the actual process of diagnosis to employ explicitly all of the cognitive processes we have been describing. Instead, the physician may simply grasp a diagnosis through the poorly understood processes of intuition or pattern-recognition. Intuitive and analytical reasoning may be complementary, Bradley suggests. For instance, hypothesis generation may be largely intuitive, and the hypotheses generated may then be analytically tested. How intuition allows a physician to "simply grasp" a diagnosis, however, is an issue that needs more explication.

Rudolf Gross suggests that intuition is a restructuring of the short- and long-term memory. In intuition, some stimulus causes a "modification of information arrangement" and results in a comparatively more favorable pattern in terms of the problem at hand. The stimulus, however, must be a fortuitous one, for any conscious effort would only result in the intensification of already existing patterns. Gross sees this form of intuition as playing a "decisive role" in diagnosis (Gross, 1993, p. 188). What I would like to suggest, however, is that intuition in the diagnostic process is not an unanalyzable form of thinking based on a fortuitous occurrence; rather, it is an expedited form of analytical reasoning that can be carried out by an expert diagnostician. Thus, intuition and expertise are intimately related.

Stuart Spicker (1993, p. 201) has distinguished two senses of intuition. In the first sense, "intuition" refers to a hunch or a guess, that is, "a belief not preceded by any inferential process." The second sense is Aristotle's notion of intuition as "immediate knowledge of the truth of a premise, though devoid of any inferential process." As Aristotle himself puts it in the *Posterior Analytics*,

> Now of the thinking states by which we grasp truth, some are unfailingly true, others admit of error – opinion, for instance, and calculation, whereas scientific knowing and intuition are always true ... and since except intuition nothing can be truer than scientific knowledge, it will be intuition that apprehends the primary premises. ... If, therefore, it is the only other kind of true thinking except scientific knowing, intuition will be the originative source of scientific knowledge (Aristotle, *Analytica Posteriora*, Bk. 2, Ch. 19, 100b 5-16, trans. Mure).

Thus Aristotle makes intuition more than a mere guess. It is a form of knowledge that seems odd to the modern scientist and philosopher, however. As Spicker (1993, p. 201) points out, the modern paradigm of knowledge has been that knowledge is *inferential*. Aristotle's notion of intuition remains a mysterious form of knowledge that cannot be explained according to any current epistemological standards.

Spicker is on the right track in his assimilation of intuition to what Michael Polanyi (1969, p. 140) has called "tacit knowing." Such knowledge is often exhibited in the performance of a skill that can be carried out competently, but for which one cannot offer an adequate explicit explanation of *how* it is carried out. Thus, intuition's role in the diagnostic process is probably less an immediate grasp of medical reality than an admission of the inadequacy of our grasp of the ways we explicitly come to know diagnoses. Intuitive beliefs may be "indispensable" in the diagnostic process, but only as "a valid anticipation of the yet indeterminate implications of the discovery or the more explicit knowledge to be arrived at in the end" (Spicker, 1993, p. 205). Intuition in diagnosis should probably be understood as a case of inferential knowledge in which the diagnostician is not explicitly aware of the particular stages of the reasoning process (Moseley, 1993, p. 215).

Such intuitions ought not to be the normative foundation for diagnosis. It is expertise that gives rise to the ability to properly use intuitions in diagnosis. Therefore, understanding expertise may be the key to understanding how physicians use intuition to come to correct diagnoses.

Expertise

Paul Johnson (1983, pp. 78-79) defines an expert as one who, because of training and experience, is able to do things the rest of us cannot do; experts are not only proficient, but also efficient. Expertise seems to be acquired in three stages. In the first stage, the individual must learn from instruction or observation what actions are appropriate in particular circumstances. In the second stage, the relationships learned in the first stage are practiced until they become accurate and efficient. In the third stage, the relationships are so well practiced that they become "automatic" and are generally not available to conscious awareness (Johnson, 1983, pp. 78-79). Thus, experts do what they do very well, but cannot completely explain how they do what they do. What this means is that the process that experts use in diagnosis cannot be described directly and formally, but must be "reconstructed."

Several empirical studies of expert diagnosticians have revealed consistent results about hypothesis generation. This is precisely the level of the diagnostic process in which intuition is likely to play its greatest role. As we have seen, clinicians often report that hypotheses just seem to "pop into their heads." On the other hand, the testing of the hypotheses is much more apt to proceed according to a process that is well enough defined to allow a systematic analysis.

Elstein, Shulman, and Sprafka (1978, p. 276) conclude that the differences between experts and weaker problem solvers depend more on the extent of experience organized in long-term memory than on differences in the problem-solving heuristics they employed. Elke Weber et al. (1993, pp. 1157-1159) found that experienced physicians were more likely than novices to generate multiple instances of the same general hypothesis in a general-to-specific order. In addition, past experience with a particular diagnosis made the expert physicians much more likely to raise the possibility in a hypothesis again. Physicians with more experience were more likely to report that they had seen a similar case before. Expert diagnosticians simply have a larger stock of hypotheses, stored in long-term memory, from which they can draw.

In their study of neurologists, Barrows and Bennett (1972, p. 276) found that, compared with students who tend to formulate quite specific hypotheses, experienced neurologists tend to keep their initial hypotheses quite broad, gradually making them more specific as more data became available. For example, a medical student seeing a patient with a sudden onset of hemiparesis might immediately think that the patient had a stroke due to cerebral vascular disease. The more experienced clinician would formulate a more general hypothesis, such as "lesion in the opposite hemisphere," and gradually refine the hypothesis by deliberately seeking data that would confirm or deny the hypothesis. The experienced clinician will also make early decisions about the urgency of the problem. On this last point, note that urgency of a problem is related to value judgments about just how disvalued the problem is.

A study of internists by James Sisson and colleages (1991, pp. 607-612) produced similar findings. Compared with physicians, medical students generated significantly more hypotheses, and significantly more specific hypotheses.

These studies suggest that experienced physicians tend to leave more options open to themselves at the beginning stages of the diagnostic process. Experts also appear to use their past experiences to pick out what Bradley (1993, p. 57) has called "pivotal features," which "function like beacon lights to guide one through this fog of information." The ability to pick out important pivotal features in a present case because of past experience allows the experienced physician to come to earlier conclusions. The way knowledge is structured in memory is as important as the depth of knowledge. Experience seems to allow experts to make better estimates of rates of disease in the population (Bradley, 1993, p.

59). This is important, as we have seen, for interpreting the predictive value of diagnostic tests.

One potential problem the above analysis brings out has to do with scope of expertise. The tendency to think that consensus of experts shows "rightness" is one part of the fallacy that Robert Veatch (1973a, pp. 29-40) has called "generalization of expertise." Experts in one area may be recognized as authorities in other areas in which they in fact have no expertise. If my argument up until now has been correct, value judgments are inherently a part of diagnosis. However, this is not very often overtly acknowledged in analyses of diagnostic reasoning. A physician who is recognized as an expert diagnostician may have no training at all in matters of values, and may even deny that diagnosis is a value-laden endeavor.

This leaves us in a quandary. It seems that we must say that no one is an expert diagnostician without expertise in the realm of values. Yet few of the people we recognize as expert diagnosticians do have such expertise in value decisions. I do not think that we should deny that expertise in diagnosis exists. The best solution to this problem may be to recognize that some diagnosticians may make good value decisions without specific training in the normative realm. This is not implausible, for most of the good people in the world know nothing whatsoever about ethical theory. However, it should serve to caution us against readily accepting appeals to the intuition of experts in diagnosis, where intuition is understood as a shortcut in diagnostic reasoning. Depending on intuition in diagnosis may lead to a glossing over of already hidden but possibly important values in diagnosis. Medical school courses in medical ethics and medical decision making would do well to stress the fact that neither medical practice nor medical science is value-neutral. Becoming more aware of the value dimension in medicine will only serve to enhance the expertise of diagnosticians.

V. VALUES AND COMPUTER DIAGNOSIS

The use of computers in diagnosis is not without problems. Some might think that these problems stem from an inability of computers to handle the "art" of making value judgments. This is partly correct. However, even if all the values we have been considering cannot be precisely quantified, there still might be ways that computers could factor values into diagnostic decisions, perhaps even better than humans can. The problems of computer diagnosis do not arise from an inability of computers to handle value judgments at this rather simple level. Rather, the problems come from the difficulties that we have found in describing an adequate and complete logic of diagnosis and from the complexity of the different levels of values that are embedded in the practice of diagnosis. Many of the values of diagnosis lie completely hidden from even expert diagnosticians.

Computer programs for diagnosis have relied on three different approaches: Bayesian logic, branching logic, and "simulated" logic. Examples of the third type are the Stanford-based MYCIN program for diagnosis and treatment of infectious diseases and the Pittsburgh-based program, which started as DIALOGUE and evolved through INTERNIST-1 and INTERNIST-2 to CADUCEUS, for diagnosis in internal medicine (Schaffner, 1992, pp. 198-221). We have already considered some of the conceptual difficulties in Bayesian and branching logic approaches to diagnosis. The same difficulties apply to computer programs using such logical approaches. I do not want to consider these problems further;[6] instead, I will explore some more general issues about the use of computers in diagnosis and the ways in which values play an important role in computer diagnosis.

A Critique of the Standard View

The "Standard View" about computer diagnosis, held by most people who have written about the subject, is that while computers may be of great help in the diagnostic process, they will never replace the human diagnostician. James Mazoué (1990, pp. 560-575) has opposed the Standard View. He argues that four alleged impediments to replacing the human diagnostician with a computer are not as serious as supposed by proponents of the Standard View. In this section, I will simply present Mazoué's critique. I will respond with a defense of the standard view in the next section.

The first impediment is that computer programs do not have the technical capacity to match an expert diagnostician. Mazoué responds that there is nothing in principle that would prevent the technical improvement of computer programs to the point that they are effective and reliable diagnosticians.

The second concern focuses on semantic issues. Diagnostic algorithms can only process information that has already been screened by someone who understands the potential diagnostic significance of the information. However, Mazoué argues that information acquisition and diagnostic reasoning are logically distinct functions. Thus, there is no reason that these two functions must be combined in a single practitioner.

The third area of concern consists of an array of practical issues: cost-effectiveness, vulnerability to tampering, and the possibility of entry of "discrepant or inconsistent information." Mazoué recognizes these concerns, but does not feel that they present insurmountable obstacles.

Finally, proponents of the Standard View raise valuational concerns, citing a "valuational gap" between world and machine. Mazoué responds that the valuational gap would have to be bridged by human practitioners when it comes to selecting appropriate treatment, but that this is a separate issue from determining the pathophysiological causes of a disease.

Mazoué argues that diagnosis might be separable from medical practice. That is, diagnosis would be given over to computers, while treatment would remain the domain of the physician. He then provocatively argues that if computers can be made to be as accurate as physicians in diagnosis, and if diagnostic skill is not *itself* necessary for medical practice, then a system of teaching clinical diagnostic skill can no longer be ethically defended. Computers *ought* to take over the task.

A Defense of the Standard View

I believe that Mazoué is overly optimistic in his assessment. It may be that technical improvements will allow computers to pass a Turing test in certain limited areas. That is, computers may be able to arrive at the same diagnoses that an expert human diagnostician would make. Adequate solutions to the practical problems such as confidentiality and tampering might also be devised. However, semantic and valuational problems would still remain.

The Problem of Validation
The deeper conceptual problems that we have unearthed in our explorations of the concept and classification of disease remain problems that are presently and perhaps forever beyond the capabilities of computers. First, the problem of validation remains unsolved. How are

we to judge whether a computer program is safe for use in diagnosis? There is no accepted method for making such a validation (Miller, Schaffner, and Meisel, 1985, pp. 534-535). Individual judgments about safety may be idiosyncratic. Furthermore, the knowledge base that goes into the computer will necessarily be an interpreted knowledge base. Some human computer programmer must make value judgments at the foundational, conceptual and nosological levels in determining what knowledge the computer will consider in the diagnostic process. Should a way to incorporate values into a computer program be found, there would have to be value judgments about which values to include. To complicate matters even further, medical knowledge is not static (Miller, Schaffner, and Meisel, 1985, pp. 534-535). Deciding the current standard of knowledge is not easy, and is probably beyond the capabilities of a computer. What all this means is that computers might become useful tools for the human diagnostician, but will not replace the human diagnostician.

Semantic Problems and Problems of Understanding
The semantic differences between human and computer are greater than Mazoué will admit. As Randolph Miller has recognized, eliciting the circumstances of the patient and the patient's illness cannot be performed by an automaton. It requires a sequence of interdependent and often highly individualized processes (Miller, 1990, pp. 583-584). Computers will generally not be able to consider the unique situation of an individual patient. Miller, Schaffner and Meisel's (1985, p. 535) example is apt: recommending coronary bypass surgery to a 45-year-old year old man disabled by angina is different from recommending the surgery to an 85-year-old patient with the same symptoms but with terminal cancer as well. Although this example concerns treatment and not diagnosis, similar considerations apply in the area of diagnosis. The considerations for recommending coronary angiography simply to gain information ought to be different in these two patients. What one will do with the information gained is an important element that must be considered in deciding whether the information ought to be gained. Thus, decisions about diagnosis can be as individual and unique as decisions about treatment.

Translating subjective findings into the precise language required by the computer is another problem. Suppose a patient smiles while reporting, "I am having the worst pain I have ever had." It is extremely difficult to know how to enter such a report into a computer program (Miller, Schaffner and Meisel, 1985, p. 535).

A related problem concerns the classification of diseases according to well-established nosologies such as SNOP, SNOMED, and ICD-9. As Peter Hucklenbroich (1988, pp. 174-176) argues, such classification systems are constructed for the purpose of uniform data storage and retrieval purposes. Their sequential, linear syntax may not be sufficiently rich for the needs of the knowledge processing of computer systems. Diagnoses may sometimes be desirable for reasons other than the reasons for which the nososlogies were constructed.

As Edmond Murphy (1992, p. 265) has observed, obsession with formal proof may even have stifled the role of meaning in discussions of rational conviction. While in principle it may be possible to include such factors of meaning in a computer diagnosis program, in practice such factors will often lack the precise formulation that will be necessary to keep the program practically useful.

Marx Wartofsky (1986, pp. 88-89) argues that expert computer programs cannot substitute for a human diagnostician because the act of diagnostic insight requires as a necessary condition an extraordinarily complex structure of background knowledge and intuition. These elements allow the human diagnostician to "catch on" to a diagnosis in the same way one "catches on" to a joke. The only hope for the computer is to be able to perform computational procedures on data and to arrive at a diagnostic conclusion that is extensionally equivalent to the one of the expert human diagnostician. Wartofsky argues that this computational approach to diagnosis leads to a view in which the method that is constructed to perform the diagnostic task "ontologizes itself." Then, the human mind becomes simply a computational machine that performs like a computer. As Wartofsky puts it, "ontology recapitulates methodology." He argues that a diagnostician does not perform adequately without the ability to "catch on" to a diagnosis, something a computer cannot do. A human diagnostician can bring to the diagnostic process a wealth of imaginative connections from any of the fields of human thought or action or feeling with which the diagnostician has experience. This is the sort of stuff that allows the human to "catch on." However, this sort of background knowledge is so complex that it is hard to conceive of programming a computer to be able to draw on such a knowledge base. Thus, a computer will never be able to grasp the unexpected combinations of data that go into making a diagnosis (Wartofsky, 1986, p. 91).

While Wartofsky is right in his analysis of the complexity of background knowledge that goes into making many diagnoses, his analogy to catching on to a joke is not entirely apt. Catching on to a joke seems to be more of an emotional response. Wartofsky is right, of course, that there is a cognitive aspect to catching on to a joke. This cognitive aspect involves understandings of cultural context and how the individual case harmonizes or fails to harmonize with that context. The complexity of this cognitive aspect may indeed be a challenge to a computer. However, given an adequate database of background knowledge, it is conceivable that a computer could be programmed to make such connections. The computer would still, of course, be unable to mount an emotional response, and in this sense would be unable to catch on to a joke.

What is really at issue here is whether some sort of *understanding* is necessary in the diagnostic process.[7] The point, then, is similar to the one made by John Searle in his "Chinese room" argument that computer programs merely manipulate symbols, while human consciousness involves an apprehension of their meaning (Searle, 1991, pp. 276-285). The question is, Is extensional equivalence of output all that is necessary for diagnosis, or is some important intensional aspect of disease omitted when computers make diagnoses? In some simple cases of diagnosis, those in which fact and value judgments are not in dispute, extensional equivalence may lead to results that will be satisfactory to all. However, a patient's experience in many cases of illness will be of such complexity that human understanding will be necessary to interpret it in order to make a diagnosis.

The Multiple Activities of Diagnosis

Another problem with Mazoué's account lies in his separation of the diagnostic and therapeutic roles of the physician. As we have seen, there is an intimate bond between diagnosis and therapy. Diagnosis in clinical medicine is not an end in itself, but a warrant for treatment or some other goal. The levels of treatment desired or the other goals desired affect how far the diagnostic process is carried, and how much explanatory power is sought. Thus, the diagnostic and therapeutic roles cannot be totally divorced.

In addition, definitive diagnoses are not always finalized before treatment is begun. Responses to particular therapies can provide important new data for further refining a preliminary diagnosis. Under such complex conditions, it is hard to take a computer system that can manipulate data without understanding the conceptual complexity of them to be semantically adequate for the important task of diagnosis (Miller, 1990, p. 584).

A large part of the difficulty in formalizing the diagnostic process to the point where computers can take over as diagnosticians lies in the fact that the diagnostic process is itself not well defined. Rather, it probably consists of a number of different activities. Marsden Blois has suggested several: the application of a conventional disease label to a patient, the detection of a causal factor for a particular malfunctioning part of the patient, and the identification of what is presumed to be a proximate *exogenous* cause of an illness. In addition, sometimes the diagnostic process is stopped before the above goals are reached because of a state of incompleteness in medical knowledge. Because of such ambiguities, there may be no formal end-point of the diagnostic process. These diagnostic activities are not disjunctive, nor are they exhaustive. Such vagueness militates against attempts to formalize diagnosis by computer (Blois 1983, pp. 30-31).

We have already seen how the physician's conception of the patient's good, as reflected in cost-benefit assessments, determines where cut-off points between abnormal and normal tests may be drawn. That is, the cost of "missing" particular diagnoses will depend on conceptions of just how serious such a mistake would be. As Edmund Pellegrino has argued, a physician's personal clinical "style," that is, the sum total of the logical and epistemological preferences that guide the physician's clinical decision making, dictates how value judgments will be brought into the diagnostic process. Some clinicians seek "elegance" in diagnosis, that is, a complete and satisfying explanation of the patient's ills. Others seek an economical approach in which a diagnosis is made using the least number of tests possible (Pellegrino, 1992, p. 181). As Pellegrino suggests, if the diagnostic process is to be seen as a hierarchical process of reasoning, it must include a moral or ethical level at its very top (Pellegrino, 1992, p. 192). This top level should also include values beyond the strictly ethical, values having to do with ultimate questions of meaning. Such values are surely an essential part of the patient's experience of illness, which is the ultimate subject matter of diagnosis.

The Aim of Diagnosis

I have maintained that diagnosis cannot be construed as simply putting particular patients into predefined nosological categories. Diagnosis is tied to treatment and prognosis; ultimately diagnosis is tied to healing, which goes beyond any strict scientific account of disease. If this is so, then it may sometimes be necessary to concede that a particular patient does not fit any established category. The diagnostic process would be more like a discovery process in which some new diagnosis is invented, tentative though it may be. In cases such as this, a computer program will not be able to properly assess the data to make a new discovery. Such cases may even call for the introduction of a new conceptual schema akin to a new thought style. As Carl Hempel (1985, pp. 119-122) has argued, such novel concepts will not be available in a computer program, and it is hard to see how a computer could be programmed to discover such concepts and introduce the new vocabulary needed for them. It is even more difficult to see how a computer could be programmed to assess the values that would have to go into the judgment about when such a new schema is necessary.

The sort of diagnostic activity undertaken in a particular case may well depend upon conceptions of disease in the hypotheses being entertained. Blois finds diseases to have a hierarchical nature. They may be described at different levels of organization: atomic, molecular, cellular, physiological system, or entire patient. The description of low-level objects requires few predicates. "Electron," for example, requires little description and is relatively unambiguous. At higher levels, we begin to find "emergent properties" that frequently cannot be adequately predicted from theories and are of such complexity that they resist precise descriptions (Blois, 1983, pp. 34-35). The experience of pain, for example, cannot be adequately explained in low-level terms such as C-fiber stimulation.

Some diseases, especially those that have not yet produced any symptoms, can be adequately described using only the lower levels. However, strict scientific accounts of disease, on the lower atomic and molecular levels, have the potential to lead to the kind of reductionism that I have already discussed and rejected. The problem with such accounts, as Blois recognizes, is that patients experience illness only at higher levels. Individuals experience pain, not C-fiber stimulation.

A few diseases are understood in such detail so that it is possible to give causal accounts that bridge the various levels (Blois, 1983, pp. 37-

39). Nonetheless, physicians ordinarily operate at higher levels in diagnosis, hearing "heart murmurs" rather than simple particular sounds. Physicians do not operate in one diagnostic mode, but all computer programs based on some logic do. It is hard to envision a computer with all the complexity required to understand a patient's illness in the way that a physician can.

These problems suggest that the proper role for the computer in diagnosis may be what J. G. Scadding (1967, p. 881) has called the "library function." Computers have the ability to store large amounts of data and make the data rapidly accessible. They may play their most valuable role in suggesting possible diagnoses beyond the ones that spontaneously occur to the clinician during the stage of hypothesis formation.

Classical decision theory has been suggested as a way to incorporate value judgments into computer diagnosis. However, as Henrik Wulff (1992, pp. 249-250) points out, these theories have close ties with classical utilitarian moral philosophy. Wulff suggests that such thinking leads to a paternalistic attitude toward the patient. The four levels of values we have been considering will lie thickly embedded in any diagnostic program. For some simple diagnoses, this will not be a problem. For others, idiosyncratic values of the patient may play an important role that cannot be recognized by the computer.

VI. CONCLUSIONS

In this chapter, we have looked at some of the complexities of the diagnostic process. We have seen that there is no agreement on a single logical method that can serve as an ideal model for diagnosticians to follow. Furthermore, there is no agreement even on a precise definition of diagnosis.

Physicians do become quite competent at diagnosis, and they will sometimes attribute their competence to sound intuition. Although intuition should not be looked upon as a sort of mystical inspiration that guides competent diagnosticians, intuition does play an important role in diagnosis. Intuition is probably better seen as the sort of "tacit knowing" that characterizes the reasoning of the expert diagnostician.

I have concluded that computers are not likely to replace human diagnosticians any time soon. Moreover, replacing human diagnosticians with computers is not desirable. The facts and values of diagnosis are complex and not yet well characterized. More importantly, however, diagnosis must take into account the subjective experience of the suffering patient. It takes a compassionate and knowledgeable human physician to do that adequately.

CHAPTER 8

CONCLUSION

We began with an examination of the challenge of social constructivism to scientific realism. Scientific realism is often seen as the fundamental basis for objectivity and rationality in medical practice. Few physicians would argue that science is the *only* foundation of Western medicine; most would recognize that medical practice is a value-laden enterprise. However, most physicians tend to separate the factual basis of medicine from the value judgments involved. I have argued that scientific realism is an inadequate philosophical basis for understanding even what seems to be a purely scientific aspect of medicine, the practice of diagnosis. Instead, I have proposed a new form of realism, value-dependent realism, which attempts to mediate between social constructivism and scientific realism.

Scientific realists take scientific theories to be approximately true descriptions of the way the world actually is. They point to the instrumental success of certain theories in explaining and predicting natural phenomena as the best evidence for scientific realism. Value-dependent realism tries to take this instrumental success seriously, but denies that our present theories, no matter how empirically adequate, are the only possible descriptions of the way the world is. Those holding different worldviews may describe the same bit of reality in radically different, even incommensurable terms, yet both descriptions may be true, or at least not obviously false. Furthermore, each description may be empirically adequate for its own particular worldview.

Value-dependent realism is a metaphysical theory about the nature of disease and diagnosis. It is open to the realistic sense that diseases exist independently of our particular theorizing about them. Our scientific theorizing about disease is never a matter of philosophizing in the quiet of our studies, our libraries, or even our laboratories. The world offers resistance to our theorizing. We cannot simply construct theories about disease according to any values or goals we choose. The hard fact is that human beings get sick and they die. We can choose to describe these phenomena in different ways. We can even adopt a spiritual worldview that denies the importance of the physical body. But we cannot avoid the empirical realities of sickness and death. People suffer from illnesses that

192

are real enough to them and are made manifest to others. Disease, which is a concept that serves to explain these illnesses, reflects this reality whether it is understood in a scientific framework or not.

We cannot, however, assume that the scientific constructs of the scientific realist are the only real explanation of illness. Our descriptions of disease necessarily use concepts, which are human constructions. It is impossible to state a fact that does not depend on some conceptual framework. Not every conceptual framework will pass the test of empirical adequacy, but there may be more than one adequate framework for stating any fact. To choose a conceptual framework is an evaluative act. All descriptions of disease, even scientific descriptions, depend on some conceptual framework imbued with the various sorts of goals and values we have been considering throughout this book.

Some descriptions of disease are better than others are. Judgments of which descriptions are better depend on which particular values we take to be most important for our description. This need not drive us into relativism, however. There is a strong case to be made for value realism, the view that certain values are objective values and ought to be held by everyone. If some values are objective values, then we need not worry that our insistence about facts being value-dependent destroys the objectivity of our scientific facts. The value-dependent realist can hold that scientific facts are objective just as strongly as does the scientific realist. But the value-dependent realist recognizes that agreement about facts depends on agreement about the underlying values.

Agreement on values does not entail agreement on facts. Groups of scientists may agree on all the relevant values in a research project about the cause of sudden infant death syndrome; they may still disagree about the facts. However, when everyone agrees that something is a fact, it is because they all agree on the values embedded in the fact. The most uncontested and seemingly incontrovertible facts are objective facts because they rest on values that are undisputed, often so undisputed that they are not even noticed.

The objective values that inform the debate about disease are the conceptual and nosological values we discussed in chapters 4 and 5. Death, suffering and disability are examples of objective disvalues that play important roles in our decisions about what constitutes a disease.[1] There are, of course, extraordinary circumstances when one rightly sees an objective disvalue as an instrumental value, such as when one accepts the suffering that accompanies some treatment that can postpone the disvalue of death. This, however, does not cast doubt on the fact that suffering remains a disvalue. People would still avoid the suffering if they could.[2]

In addition, we may agree on the fact that disability is a disvalue, but disagree about whether a particular condition is a disability. While most people have tended to see deafness as a disability, many in the deaf community deny this, holding that deafness merely represents a difference that binds a sociocultural community with a distinct means of communication.[3] Participants in this dispute might agree on the foundational values that permit all parties to agree on what constitutes deafness, but disagree on the conceptual values that hold deafness to be a biomedical problem. Disagreement does not entail that the value is only a matter of personal taste. One side might just be wrong in its assessment of the values and disvalues involved in calling deafness a biomedical problem. This does not present problems for the value-dependent realist any more than disagreements about the facts about, say, genetic screening for breast cancer[4] present problems for the scientific realist. The value-dependent realist must present persuasive evidence for value claims just as the scientific realist must present persuasive evidence for claims of scientific fact. The goal for value-dependent realists is to present enough persuasive evidence to convince all that the values they claim to be objective are indeed objective.

The focus of this work has been on a particular facet of medical practice: diagnosis. Diagnosis deals with facts. Diagnosis is an attempt to understand the nature of biomedical problems. We generally carry out the diagnostic process because we take it to provide a rational warrant for treatment and for making prognoses. There may be other reasons why we want to label people as having certain categories of disease. Whatever the reason, diagnosis is an essential part of medical practice, and depends not only on the facts that can be ascertained about a person who is ill, but also on our concepts of disease and the nosologies by which we classify diseases.

I have argued that the concept of disease, the classification of diseases, and the process of diagnosis are best understood within a framework of value-dependent realism. In this framework, values enter diagnosis at four levels. The first is the level of our understanding of what constitutes a fact, and of our choice of a framework for describing facts. All our scientific facts, including medical facts about disease, are value-laden. These sorts of values are what I have called foundational values.

The heart of science is theory formulation and empirical testing and measurement. Foundational values are involved in the selection of theories and the establishment of standards for measurement and precision. Such basic foundational values as truth, accuracy, simplicity, and predictability are involved in setting the standards for deciding what we take to be facts. Epistemic values are important foundational values. We expect harmony of the components of a theory. For scientific theories, we expect a high degree of empirical content. Theories should be fruitful in making predictions, and have heuristic power. In science, hypotheses must be thoroughly tested before they are said to tell us the facts. This process depends on the sociocultural and professional values of the relevant scientific community that decides the adequacy of hypothesis testing. All of these are foundational value judgments about what a scientific theory should be, and what is necessary to consider some claim as stating a fact.

Diagnosis is ostensibly about discovering the nature of a disease. I have argued that the concept of disease is essentially value-laden with what I have called conceptual values. Non-normativists think that disease can be defined without reference to value terms. To do this, they make use of the notion of natural functions, often relying on evolutionary theory to provide the basis to say which functions are natural. However, as the normativists have shown, attempts to reduce the goals of an organism to such terms without reference to values fail. The choice of goals to specify proper functioning is itself a value judgment.

Choice of goals is characteristic of the human being. One chooses goals in accord with particular values. Both individual values and sociocultural values may come into play. These values determine what we consider to constitute a threat, what is a harm, and what we would like to be able to control about our physiological natures. Some of these values are universally held. Such things as death, loss of limb, loss of vision or hearing, and intractable pain are, under ordinary circumstances, universally held as disvalues. This agreement on values is what makes it appear to some that the concept of disease can be understood without reference to value. Other conditions, however, may appear as diseases for some but not for others. The goals of the philosopher, the pianist, and the professional athlete make very different demands on the body. What one sees as a disease may not bother the other two. Thus, some biomedical problems are universal and some are particular. This need not trouble value-dependent realists, for they make no claim that all conceptual values about disease are universal, but only that some values, under the usual circumstances as discussed above, are universal.

One can also conceptualize disease according to different metaphysical frameworks. We examined two basic conceptions of diseases: ontological and physiological. Ontological conceptions of disease take disease to be either an entity such as a humor that invades the body from without, or an entity internal to the body, such as a group of cells gone awry to form a cancer. Physiological conceptions of disease rely on some understanding of normal functioning of the body and see disease as a deviation from that norm. Different physiological conceptions of disease that have been proposed include disease as humor imbalance, disease as energy imbalance, disease as imbalance of homeostasis, diseases as temporal construction, and diseases as spatial construction. Certain of these metaphysical conceptions of disease are more useful than others are for describing particular diseases. In diagnosis, sometimes a temporal understanding of disease is essential; at other times, a spatial understanding is more useful. Our metaphysical conception of disease in any particular case will reflect conceptual normative judgments not only about the disvalued nature of the condition, but also about which particular metaphysical view we favor. Simple factual descriptions of these conditions in the terms of science alone omit an essential aspect of disease: the values that are held by the individual who suffers from the disease, which may be objective or subjective, and which help to constitute that condition as a disease for the individual.

There is no natural kind that is described by the term "disease" even though there may be naturally occurring clusters of phenomena that we name as particular diseases. Even if there are naturally occurring clusters of phenomena that we call diseases, however, there is more than one way to construct a classification of all diseases. The problem of constructing a nosology stems not only from the fact that there are different metaphysical conceptions of disease that must be included, but also from the fact that other conditions that are not properly considered diseases, but do fall into the biomedical realm, must be included. Injuries, for example, are usually considered to be distinct from diseases, but are often brought to the physician for diagnosis and treatment. How to best classify this disparate group of biomedical problems requires what I have called nosological values. Ideally, we want our nosologies to be exhaustive, useful, as simple as possible, and to contain disjoint categories. In any exhaustive nosology, however, simplicity and disjointness must be sacrificed in some way. Which parts are sacrificed, and the extent to which they are sacrificed will depend on value judgments about what is considered to be most important about the conditions in question and what is most useful for whatever goals are chosen.

Diseases are classified in a variety of ways: descriptively according to symptoms, anatomically, according to causative factors, or according to the simultaneous occurrence of several characteristics. Even a particular disease, e.g., poliomyelitis, is often classified in more than one way. Symptom diagnoses are generally not considered to be definitive until they have acquired some causal explanation. Nonetheless, symptom diagnoses serve a useful purpose in steering further investigations. Sometimes it is most important to classify a particular biomedical problem as an anatomically defined condition. This is especially true for many injuries. Causal classifications are most useful in providing a rational basis for treatment of infectious diseases and disorders of particular physiological functions. The type of biomedical problem that most obviously exhibits the nosological values that go into disease classification is the syndrome. Syndromes are often simply descriptions of clusters of phenomena that await an explanation for the clustering of those particular phenomena. We considered rheumatoid arthritis, Cushing's Syndrome, AIDS, and SIDS as examples. We saw how the very definitions of the syndromes have evolved over time as a function of new discoveries about etiology, changing social values, and even recognition of conceptual problems with the definition of the syndrome as a particular disease.

Diagnoses are constructed from facts gleaned by the physician from the history of the patient's illness, the physical examination, and various laboratory and other diagnostic tests. At this stage of diagnosis, the values involved in the construction process are less covert. I have called these values diagnostic values. Epistemic and aesthetic values are involved in the conceptual processes of recognizing the signs of disease and in separating relevant information from background noise. In addition, moral, sociocultural, and economic values are involved in diagnosis because patients are subjected to various risks and economic costs from diagnostic procedures. The benefits of obtaining a diagnosis must be weighed against these risks and costs.

Both patient and physician make contributions to the history that is written in the medical record. This written history turns out to be a narrative that is somewhat different from the narrative that the patient originally tells. The patient's telling and the diagnostician's retelling of the narrative may in fact be partially incommensurable accounts. In turning a patient's narrative into a medical narrative, the physician cannot help making new contributions to the history.

In the physical examination, the physician must necessarily interpret what he or she observes about the patient. Psychological factors can have a great influence on what is observed. Even in the absence of distortions in observation, the observer must make decisions about what observations are relevant to the diagnostic process. What is taken to be relevant will depend on the conceptual scheme that the examiner holds to be relevant to the diagnostic possibilities being entertained. As we have seen, the choice of conceptual schemes itself involves value judgments.

Diagnostic tests are used in an attempt to mitigate the uncertainties inherent in the observations, which are sometimes ambiguous, of the physical examination. Value judgments enter into several phases of diagnostic testing as well. The choice of which tests to perform will depend on a complex value calculation that considers the various costs of the tests, and the value of the information that is likely to be obtained. Once a test is performed, the results must be interpreted. The conceptual commitments that enter into such matters as choosing a normal range will include a normative dimension.

A diagnosis depends not only on an individual patient and the physician who is involved in the diagnostic process, but also on a conceptual system, a diagnostic method, a set of data, and a particular time at which the diagnosis is made. We have seen how values are inherent in the conceptual system, and in choosing the data that are considered in the diagnostic process. Further value judgments must be made in deciding upon a diagnostic process. Studies in the logic of diagnosis have shown that there is no one system of formal logic that adequately captures the method that physicians actually use to make diagnoses. Physicians must decide how to proceed in individual cases. Different diagnosticians may have different styles of diagnosis, styles which all may lead to a reasonable outcome but do so by different routes.

Another part of the problem of formalizing the logic of diagnosis lies in the vagueness of many of the terms used by both patient and physician in describing the manifestations of disease. This is one important reason why computers will not completely take over the task of diagnosis from physicians. The problem is not that computers cannot take values into consideration; if the relevant values could be incorporated into a computer program, the computer should be able to factor them into the diagnostic process as well as a human diagnostician. The problem lies in the complexity of the logical process of balancing so many vaguely defined conceptual and evaluative elements and not in the mere presence of values in the process. Nonetheless, the computer may still play an invaluable role in diagnosis as a repository of information on both fact and value.

Medicine is an art in the sense that it goes beyond an application of the rational methods of science. The art of medicine involves not only such things as tact in the patient-physician relationship, but also the evaluation of all the elements of diagnosis. Thus art and science in medicine cannot be separated. Making the four levels of values that reside in diagnosis more explicit and precise can only help to make the science of diagnosis more precise as well. When the value dimensions of disease and diagnosis are recognized, physicians are less likely to fall into the trap of diagnostic dogmatism, which is the result of an unsophisticated reliance on scientific method, and which cannot tell the whole story of disease.

NOTES

CHAPTER 1: INTRODUCTION

[1] I draw this distinction between art and science in medicine rather crudely, simply to point out the tension created by those who want to maintain the gulf between value-laden art and value-free science. There are many more sophisticated analyses of the "art" of medicine. See, for example, the discussion of Pellegrino and Thomasma (1981, pp. 144-149).

[2] Cf. Heidegger (1996), who makes a similar point, albeit from a very different approach. Heidegger argues that values are "*objectively present* determinations of a thing" (p. 92). We do not attach values to "what is nakedly objectively present." Rather, "what is encountered in the world is always already in a relevance which is disclosed in the understanding of world, a relevance which is made explicit by interpretation" (p. 140). A Heideggerian approach to the problem of diagnosis might be fruitful, but I will not further pursue it in this book.

CHAPTER 2: SOCIAL CONSTRUCTIVISM VS. SCIENTIFIC REALISM

[1] The most important of these from the standpoint of this present work is that scientific methodology, scientific concepts and terms of scientific language are all theory-dependent. Thus, any simple verificationist conception of science is no longer tenable. See Boyd, Gasper, and Trout (1991, pp. xi-xiii) for this and several other points of consensus.

[2] Trenn's article is a translation of Fleck (1935) with a short introduction by Trenn. Fleck's article is a précis of his *Genesis and Development of a Scientific Fact* (1979).

[3] The following analysis of the criticisms of Kuhn's position is adapted from Veatch and Stempsey (1995).

[4] For an extensive listing of such criticisms, see Hoyningen-Huene (1993, p. 207, n. 58).

[5] For a discussion of the influence of Fleck, as well as Wittgenstein and Quine, on Kuhn, see Cedarbaum (1983, pp. 173-213).

[6] Steve Fuller (1988, pp. 85-89) calls Kuhn's position "many-worlds realism." That is, he takes Kuhn to believe in the independent reality of all scientifically possible worlds (i.e., paradigms) except the actual one, whose reality depends entirely on what is negotiated by the relevant group of scientists.

[7] The terms "local" and "global" realism are favored by Vision (1988, pp. 4-6).

[8] The social constructivist would be unlikely to say that scientific facts are subjective. However, for the true constructivist, objectivity amounts to the subjective agreement of those within the relevant community.

CHAPTER 3: FACT VS. VALUE

[1] This is a reprint of an article appearing in *Proceedings of the Aristotelian Society* 24 supp. (1950).

[2] Strawson uses "statement" and not "proposition." His use of "statement," however, is somewhat ambiguous. It appears that Strawson does not mean to imply that a fact must actually be stated in order to be fact. Thus, his usage of "statement" is apparently equivalent to the common usage of "proposition." See Shorter, (1962, pp. 295-296).

[3] For a concise discussion of these groundings of value, see Rescher (1969, pp. 49-53).

[4] Thomas Nagel (1986, pp. 143-149) argues that in the controversy about value realism the burden of proof has been misplaced, and should rest on anti-realists rather than realists. Realism about values, although defeasible, is entirely reasonable. I think Nagel is right about this, although I do not believe we can discover these values by seeking out a "view from nowhere." Everybody has to be somewhere.

[5] See, for example, Rawls (1971).

[6] For recent examples of objectivist naturalism, see Railton (1986b) and Boyd (1988).

[7] See Rawls (1971, pp. 48-51) for an explanation of the concept of reflective equilibrium.

[8] See Stack (1969) and Stack (1973).

[9] See, for example, Railton (1986b, pp. 182-184).

[10] All the examples given are from Rescher (1969, pp. 14-19).

[11] Examples of models are clinical, engineering, collegial, and contractual. All of these have some application in the realm of medicine.

[12] See Tumulty (1988, pp. 53-63).

[13] This point was suggested to me by Robert Veatch.

CHAPTER 4: THE CONCEPT OF DISEASE

[1] See, for example, Temkin (1973, p. 402), and Engelhardt, (1975, p. 130).

[2] Excerpts from this and other works of members of the Polish school have been translated into English and can be found in Löwy (1990b).

[3] Engelhardt (1974) offers a similar conception of disease. He reads Virchow's later work as putting forth this sort of bridge between ontological and physiological conceptions of disease. This is, he admits, a heuristic and not a strict reading. I do not wish to enter into exegetical quibbles about Virchow here. I do, however, believe that my account agrees in substance with Engelhardt's.

[4] Whitbeck (1978) gives parallel definitions of injury and impairment. I will discuss the classification of diseases and the difference between diseases and other medical conditions in the next chapter.

[5] This view has the consequence that health is not a univocal concept, but is essentially relative to the goals of the individual. This should not bother us too much if we consider that many human goals are indeed universal or near universal goals. For example, breathing is a goal that is fundamental and necessary for the fulfillment of virtually all other goals. Hence, we can agree, in large part, that a process that disrupts the ability to breathe does constitute a diminishment in health to virtually everyone. For another

treatment of health and disease that is in fundamental agreement with this point, see Nordenfelt (1997, pp. 77-78).

[6] Brown follows Whitbeck in seeing disease as a "psycho-physiological" process. I am not so interested in retaining psychological disorders in the realm of disease. However, the point that the clause can easily be shown to have value content stands.

CHAPTER 5: THE CLASSIFICATION OF DISEASES

[1] Knud Faber (1923) traces the most important developments in nosology from the seventeenth up to the early twentieth century. I depend on this work and on Engelhardt (1985, pp. 56-71) for information about historical nosologies.

[2] The description that follows is adapted from the introduction to *Systematized Nomenclature of Pathology* (College of American Pathologists, Committee on Nomenclature and Classification of Disease, 1965, pp. xiv-xviii).

[3] I take neoplasm to describe a function, that is, the behavior of a particular group of cells. This seems more plausible than taking neoplasms to be anatomical structures because the neoplasm category includes neoplasms from many different anatomical sites. Were the anatomic site taken to be primary, neoplasms would be classified with the diseases of particular anatomic sites. In fact, the heterogeneity of the etiology, epidemiology, and morphology of the processes known collectively as "cancer" suggests that no precise classification of cancer is possible, and that at most a Wittgensteinian family resemblance holds. See Vineis (1993, pp. 249-256).

[4] For descriptions of understandings of tuberculosis in literature, see Sonntag (1979).

[5] Ziporyn makes a similar point about syndromes, which she considers not to be "full-fledged diseases." See Ziporyn (1992, p. 118).

[6] I cannot consider the complexities of the concept of causation here, but see Mackie (1974) and Sosa and Tooley (1993).

[7] This is not true, strictly speaking. I do not mean to introduce the liar paradox here, but only to point out that the diagnosis is usually made from clinical evidence of pneumonia and indirect evidence (by culture of sputum samples, for instance) of the bacterium in the lung.

[8] See Mackie (1974, p. 62). "Inus" is shorthand for "*i*nsufficient but *n*on-redundant part of an *u*nnecessary but *s*ufficient condition. This is, for all practical purposes, the same as an "insufficient but necessary part of an unnecessary but sufficient condition."

[9] The set of seven criteria as stated is called the traditional format. A branching decision tree containing five criteria, the traditional seven minus morning stiffness and rheumatoid nodules, is also offered and said to be slightly more accurate than the traditional format in classifying patients as having rheumatoid arthritis. Both methods, however, are approved, although it is recommended that the method used be reported. For purposes of the present discussion, the difference between the two methods does not matter, for what is at issue is only the fact that certain specified criteria are being used to classify a patient as having a particular disease.

[10] For a history of the naming of AIDS, see Grmek (1990, pp. 32-33).

[11] Such diseases included, in addition to Kaposi's sarcoma and *Pneumocystis carinii* pneumonia, "apergillosis, candidiasis, cryptococcosis, cytomegalovirus, nocardiosis, strongyloidosis, toxoplasmosis, zygomycosis, or atypical mycobacteriosis (species other

than tuberculosis or lepra); esophagitis due to candidiasis, cytomegalovirus, or herpes simplex virus; progressive multifocal leukoencephalopathy; chronic enterocolitis (more than four weeks) due to cryptosporidiosis; or unusually extensive mucocutaneous herpes simplex of more than five weeks duration."

[12] For a discussion of the logical concepts involved in disease classification, see Taylor (1979, pp. 46-59) and Taylor (1981, pp. 277-286).

[13] Presentation of J. Huber, in Jordan and Kopp (1994, pp. 272-273).

CHAPTER 6: THE ELEMENTS OF DIAGNOSIS

[1] See the discussion on ontological conceptions of disease in Foucault (1975), chapter 4. Although Foucault discusses disease as primarily spatial, he also lays stress on the fact that the concept of disease is historically (and thus temporally) conditioned.

[2] See Schwartz and Griffin (1986, pp. 35-37) for a discussion of several studies.

[3] For a discussion of such implications see Nelkin and Tancredi (1989).

[4] See Statland, et al. (1979, pp. 525-533) for a discussion of these concepts.

CHAPTER 7: THE PROCESS OF DIAGNOSIS

[1] For a philosophical overview of the concept of vagueness, see Keefe and Smith (1996, pp. 1-57).

[2] For a simple proof of Bayes's Theorem, see Bradley (1993, pp. 89-90).

[3] See Schaffner (1981, pp. 168-69) for this more complex form.

[4] A similar point is made by Murphy (1988, pp.159-60).

[5] For a summary of the results of this study, see Elstein, Shulman, and Sprafka (1978, pp. 275-81).

[6] For additional discussion of these computer programs, see Schaffner (1981, pp.163-199) and Schaffner's introductory chapter in Schaffner (1985, pp. 1-32).

[7] For a discussion of philosophical problems in artificial intelligence, see Dreyfus and Dreyfus (1986).

CHAPTER 8: CONCLUSION

[1] See Clouser, Culver, and Gert (1981, p. 31) for an argument that these are objective disvalues. They would agree with me, I believe, that such objective disvalues can be instrumentally valued in some circumstances.

[2] Some might find incoherent the idea that disvalues might be valuable in certain circumstances. Why not just say that suffering, for example, is sometimes valuable? I think that this move would lead us too far toward an outright relativism. Some disvalues are purely personal and idiosyncratic. Just because I disvalue anchovies on pizza does not mean that everyone ought to disvalue anchovies on pizza. Other disvalues, such as suffering, seem to be universally held as disvalues, or at least *ought* to be universally held

as disvalues. We rightly look askance at people who say they value suffering *for its own sake*. Values and disvalues such as this are what I refer to as objective values and disvalues. They are not dependent on personal and idiosyncratic preferences.

[3]   Those who see deafness not as a disability but as an identifying mark of a cultural community refer to this community as Deaf (with a capital D). See Lane (1994), especially pp. 13-28, for a sociological account of the way relations between the Deaf and hearing communities have influenced the question of whether or not deafness is a disease.

[4]   For a discussion of such factual disagreements, see Healy (1997, pp. 1448-1449).

BIBLIOGRAPHY

Agassi, J.: 1976, 'Causality and medicine', *Journal of Medicine and Philosophy* 1, 301-317.

Agich, G. J.: 1983, 'Disease and value: A rejection of the value-neutrality thesis', *Theoretical Medicine* 4, 27-41.

Aldrich, V.: 1989, 'Photographing a fact?' *American Philosophical Quarterly* 26, 81-84.

Aristotle: 1947, *Analytica Posteriora*, in McKeon, R. (ed.), *Introduction to Aristotle*, Modern Library, New York, pp. 9-109.

Aristotle: 1985, *Nicomachean Ethics*, translated by Irwin, T., Hackett Publishing Co., Indianapolis.

Arnett, F.C., Edworthy, S.M., Bloch, D.A., McShane, D.A., Fries, J.F., Cooper, N.S., Healey, L.A., Kaplan, S.R., Liang, M.H., Luthra, H.S., Medsger, T.A., Jr., Mitchell, D.M., Neustadt, D.H., Pinals, S.R., Schaller, J.G., Sharp, J.T., Wilder, R.L., and Hunder, G.G.: 1988, 'The American Rheumatism Association 1987 revised criteria for the classification of rheumatoid arthritis', *Arthritis and Rheumatism* 31, 315-324.

Austin, J. L.: 1979, *Philosophical Papers*, 3d ed., Oxford University Press, Oxford.

Barrows, H.S., and Bennett, K.: 1972, 'The diagnostic (problem solving) skill of the neurologist: Experimental studies and their implications for neurological training', *Archives of Neurology* 26, 273-277.

Bechtel, W.: 1985, 'In defense of a naturalistic concept of health', in Humber, J.M. and Almeder, R.F. (eds.), *Biomedical Ethics Reviews—1985*, Humana Press, Clifton, New Jersey, pp. 131-170.

Beckner, M.: 1979, 'Comments on Murphy's 'Classification and its alternatives'', in Engelhardt, H.T., Jr., Spicker, S.F., and Towers, B. (eds.), *Clinical Judgment: A Critical Appraisal*, D. Reidel Publishing Co., Dordrecht, The Netherlands, pp. 87-92.

Beckwith, J.B.: 1969, 'Discussion of terminology and definition of the Sudden Infant Death Syndrome', in Bergman, A.B., Beckwith, J.B., and Ray, C.G. (eds.), *Sudden Infant Death Syndrome: Proceedings of the Second International Conference on Causes of Sudden Death in Infants*, University of Washington Press, Seattle, pp. 14-22.

Benson, E.S.: 1972, 'The concept of the normal range', *Human Pathology* 3, 152-155.

Bentham, J.: 1973, *An Introduction to the Principles of Morals and Legislation* (1798), in *The Utilitarians*, Anchor Books, New York, pp. 5-398.

Bernard, C.:, 1949, *An Introduction to the Study of Experimental Medicine*, translated by Greene, H.C., introduction by Henderson, L.J., Henry Schuman, n.p.

Blois, M.S.: 1983, 'Conceptual issues in computer-aided diagnosis and the hierarchical nature of medical knowledge', *Journal of Medicine and Philosophy* 8, 20-50.

Blois, M.S.: 1988, 'Medicine and the nature of vertical reasoning', *New England Journal of Medicine* 318, 847-851.

Boorse, C.: 1975, 'On the distinction between disease and illness', *Philosophy and Public Affairs* 5, 49-68.

Boorse, C.: 1977, 'Health as a theoretical concept', *Philosophy of Science* 44, 542-573.

Boyd, R., Gasper, P., and Trout, J. D. (eds.): 1991, *The Philosophy of Science*, MIT Press, Cambridge, Massachusetts.

Boyd, R.N.: 1988, 'How to be a moral realist', in Sayre-McCord, G. (ed.), *Essays on Moral Realism*, Cornell University Press, Ithaca, New York, pp. 181-228.

Bradley, G.W.: 1993, *Disease, Diagnosis and Decisions*, John Wiley & Sons, Chichester, England.

Brown, J.: 1804, *The Elements of Medicine*, revised and corrected with a biographical preface by Beddoes, T., William & Daniel Treadwell, Portsmouth, New Hampshire.

Brown, W.M.: 1985, 'On defining 'disease'', *Journal of Medicine and Philosophy* 10, 311-328.

Buehler, J.W. and Ward, J.W.: 1993, 'A new definition for AIDS surveillance', *Annals of Internal Medicine* 118, 390-392.

Bunzl, M.: 1980, 'Discussion: Comment on 'Health as a theoretical concept'', *Philosophy of Science* 47, 116-118.

Canguilhem, G.: 1988, *Ideology and Rationality in the History of the Life Sciences*, MIT Press, Cambridge, Massachusetts.

Canguilhem, G.: 1991, *The Normal and the Pathological*, translated by Fawcett, C.R. in collaboration with Cohen, R.S., introduction by Foucault, M., Zone Books, New York.

Cargile, J: 1996, 'The Sorites Paradox', in Keefe, R. and Smith, P. (eds.), *Vagueness: A Reader*, MIT Press, Cambridge, Massachusetts, pp. 89-98.

Carnap, R.: 1956, *Meaning and Necessity: A Study in Semantics and Modal Logic*, 2d ed., University of Chicago Press, Chicago.

Cassam, Q.: 1986, 'Science and essence', *Philosophy* 61, 95-107.

Cedarbaum, D.G.: 1983, 'Paradigms', *Studies in History and Philosophy of Science* 14, 173-213.

Center for Disease Control: 1982, 'Update on Acquired Immune Deficiency Syndrome (AIDS)—United States', *Morbidity and Mortality Weekly Report* 31, 507-514.

Centers for Disease Control: 1987, 'Revision of the CDC surveillance case definition for Acquired Immunodeficiency Syndrome', *Morbidity and Mortality Weekly Report* 36 (supp. 1S), 3S-15S.

Centers for Disease Control: 1992, '1993 Revised classification system for HIV infection and expanded surveillance case definition for AIDS among adolescents and adults', *Morbidity and Mortality Weekly Report* 41 (No. RR-17), 1-19.

Cicero: 1979, *Academica*, in Rackham, H. (trans.), *Cicero*, vol. 19, Loeb Classical Library, Harvard University Press, Cambridge, Massachusetts, pp. 397-659.

Clouser, K.D.: 1985, 'Approaching the logic of diagnosis', in Schaffner, K.F. (ed.), *Logic of Discovery and Diagnosis in Medicine*, University of California Press, Berkeley, pp. 35-55.

Clouser, K.D., Culver, C.M., and Gert, B.: 1981, 'Malady: A new treatment of disease', *Hastings Center Report* 11 (3), 29-37.

Cohen, H.: 1961, 'The evolution of the concept of disease', in Lush, B. (ed.), *Concepts of Medicine: A Collection of Essays on Aspects of Medicine*, Pergamon Press, New York, pp. 159-169.

Cohen, L.J.: 1980, 'Bayesianism versus Baconianism in the evaluation of medical diagnosis', *British Journal for the Philosophy of Science* 31, 45-62.

College of American Pathologists Committee on Nomenclature and Classification of Disease: 1965, *Systematized Nomenclature of Pathology*, College of American Pathologists, Chicago.

College of American Pathologists Committee on Nomenclature and Classification of Disease: 1979, *Systematized Nomenclature of Medicine*, 2d ed., College of American Pathologists, Skokie, Illinois

Cragg, A.W.: 1976, 'Functional words, facts and values', *Canadian Journal of Philosophy* 6, 77-94.

D'Amico, R.: 1995, 'Is disease a natural kind?', *Journal of Medicine and Philosophy* 20, 551-569.

Daniel, S.L.: 1986, 'The patient as text: A model of clinical hermeneutics, *Theoretical Medicine* 7, 195-210.

Donnellan, K.S.: 1983, 'Kripke and Putnam on natural kind terms', in Ginet, C. and Shoemaker, S. (eds.), *Knowledge and Mind: Philosophical Essays*, Oxford University Press, New York, pp. 84-104.

Doroszewski, J.: 1980, 'Hypothetico-nomological aspects of medical diagnosis', *Metamedicine* 1, 177-194.

Dreyfus, H. L. and Dreyfus, S.E.: 1986, *Mind Over Machine: The Power of Human Intuition and Expertise in the Era of the Computer*, Free Press, New York.

Dubos, R.: 1968, *Man, Medicine, and Environment*, Frederick A. Praeger, New York.

Duffin, J.M.: 1986, 'The medical philosophy of R.T.H. Laennec', *History and Philosophy of the Life Sciences* 8, 195-219.

Dupré, J.: 1981, 'Natural kinds and biological taxa', *Philosophical Review* 90, 66-90.

Elffors, L.: 1988, 'On assessing the validity of the main diagnosis in patient databases: The impact of aims for making diagnosis', *Theoretical Medicine* 9, 141-150.

Elstein, A.S., Shulman, L.S., and Sprafka, S.A.: 1978, *Medical Problem Solving: An Analysis of Clinical Reasoning*, Harvard University Press, Cambridge, Massachusetts.

Engelhardt, H.T., Jr.: 1974, 'Explanatory models in medicine: facts, theories, and values', *Texas Reports on Biology and Medicine* 32, 225-239.

Engelhardt, H.T., Jr.: 1975, 'The concepts of health and disease', in Engelhardt, H.T., Jr., and Spicker, S.F. (eds.), *Evaluation and Explanation in the Biomedical Sciences: Proceedings of the First Trans-Disciplinary Symposium on Philosophy and Medicine Held at Galveston, May 9-11, 1974*, D. Reidel Publishing Co., Dordrecht, The Netherlands, pp. 125-141.

Engelhardt, H.T., Jr.: 1976, 'Ideology and etiology', *Journal of Medicine and Philosophy* 1, 256-268.

Engelhardt, H.T., Jr.: 1980, 'Ethical issues in diagnosis', *Metamedicine* 1, 39-50.

Engelhardt, H.T., Jr.: 1981, 'Clinical judgment', *Metamedicine* 2, 301-317.

Engelhardt, H.T., Jr.: 1985, 'Typologies of disease: Nosologies revisited', in Schaffner, K.F. (ed.), *Logic of Discovery and Diagnosis in Medicine*, University of California Press, Berkeley, pp. 56-71.

Engelhardt, H.T., Jr.: 1996, *The Foundations of Bioethics*, 2d ed., Oxford University Press, New York.

Faber, K.: 1923, *Nosography in Modern Internal Medicine*, Paul B. Hoeber, New York.

Fabrega, H. Jr.: 1972, 'Concepts of disease: Logical features and social implications', *Perspectives in Biology and Medicine* 15, 583-616.

Fabrega, H. Jr.: 1979, 'The scientific usefulness of the idea of illness', *Perspectives in Biology and Medicine* 22, 545-558.

Fedoryka, K.: 1997, 'Health as a normative concept: Towards a new conceptual framework', *Journal of Medicine and Philosophy* 22, 143-160.

Feinstein, A.R.: 1967, *Clinical Judgment*, Williams & Wilkins, Baltimore.

Feinstein, A.R.: 1973a, 'An analysis of diagnostic reasoning: I. The domains and disorders of clinical macrobiology', *Yale Journal of Biology and Medicine* 46, 212-232.

Feinstein, A.R.: 1973b, 'An analysis of diagnostic reasoning: II. The strategy of intermediate decisions', *Yale Journal of Biology and Medicine* 46, 264-283.

Feinstein, A.R.: 1974, 'An analysis of diagnostic reasoning: III. The construction of clinical algorithms', *Yale Journal of Biology and Medicine* 47, 5-32.

Firth, R.: 1952, 'Ethical absolutism and the ideal observer', *Philosophy and Phenomenological Research* 12, 317-345.

Fleck, L.: 1935, 'Zur Frage der Grundlagen der medizinischen Erkenntnis', *Klinische Wochenschrift* 14, 1255-1259.

Fleck, L.: 1979, *Genesis and Development of a Scientific Fact*, edited by Trenn, T.J. and Merton, R.K., translated by Bradley, F. and Trenn, T.J., University of Chicago Press, Chicago.

Foucault, M.: 1975, *The Birth of the Clinic: An Archaeology of Medical Perception*, translated by Sheridan Smith, A.M., Vintage Books, New York.

Frankena, W.: 1980, *Ethics*, 2d ed., Prentice-Hall, Englewood Cliffs, New Jersey.

Freed, G.E., Steinschneider, A., Glassman, M., and Winn, K.: 1994, 'Sudden Infant Death Syndrome prevention and an understanding of selected clinical issues', *Pediatric Clinics of North America* 41, 967-990.

Frege, G.: 1949, 'On sense and nominatum', in Feigl, H., and Sellars, W (eds.), *Readings in Philosophical Analysis*, Appleton-Century-Crofts, New York, pp. 85-102.

Fuller, S.: 1988, *Social Epistemology*, Indiana University Press, Bloomington.

Goosens, W.K.: 1980, 'Values, health, and medicine', *Philosophy of Science* 47, 100-115.

Grmek, M.D.: 1990, *History of AIDS: Emergence and Origin of a Modern Pandemic*, translated by Maulitz R.C. and Duffin, J., Princeton University Press, Princeton, New Jersey.

Gross, F.: 1971, 'The Emperor's Clothes Syndrome', *New England Journal of Medicine* 285, 863.

Gross, R.: 1993, 'Intuition and technology as bases of medical decision-making', in Delkeskamp-Hayes, C. and Gardell Cutter, M.A. (eds.), *Science, Technology, and the Art of Medicine*, Kluwer Academic Publishers, Dordrecht, The Netherlands, pp. 183-197.

Hamlyn, D.W.: 1962, 'The correspondence theory of truth', *Philosophical Quarterly* 12, 193-205.

Handy, R.: 1960, 'A need definition of 'value'', *Philosophical Quarterly* 10, 156-163.

Hare, R.M.: 1952, *The Language of Morals*, Oxford University Press, London, reprint 1975.

Healy, B.: 1997, 'BRCA genes—Bookmaking, fortunetelling, and medical care', *New England Journal of Medicine* 336, 1448-1449.

Heidegger, M.: 1996, *Being and Time*, translated by Stambaugh, J., State University of New York Press, Albany.

Hempel, C.G.: 1965, *Aspects of Scientific Explanation and Other Essays in the Philosophy of Science*, Free Press, New York.

Hempel, C.G.: 1985, 'Thoughts on the limitations of discovery by computer', in Schaffner, K.F. (ed.), *Logic of Discovery and Diagnosis in Medicine*, University of California Press, Berkeley, pp. 115-122.

Hippocrates: 1981, *The Sacred Disease*, in Jones, W.H.S. (trans.), *Hippocrates*, vol. 2, Loeb Classical Library, Harvard University Press, Cambridge, Massachusetts, pp. 127-183.

Hippocrates: 1984, *Ancient Medicine*, in Jones, W.H.S. (trans.), *Hippocrates*, vol. 1, Loeb Classical Library, Harvard University Press, Cambridge, Massachusetts, pp. 1-64.

Hoffman, H.J. and Hillman, L.S.: 1992, 'Epidemiology of the Sudden Infant Death Syndrome: Maternal, neonatal, and postneonatal risk factors', *Clinics in Perinatology* 19, 717-737.

Holtzman, N.A.: 1994, 'The interpretation of laboratory results: The paradoxical effect of medical training', *Journal of Clinical Ethics* 2, 241-243.

Hoyningen-Huene, P.: 1990, 'Kuhn's conception of incommensurability', *Studies in History and Philosophy of Science* 21, 481-492.

Hoyningen-Huene, P.: 1993, *Reconstructing Scientific Revolutions: Thomas S. Kuhn's Philosophy of Science*, translated by Alexander T. Levine, with a forward by Thomas S. Kuhn, University of Chicago Press, Chicago.

Hucklenbroich, P.: 1988, 'Problems of nomenclature and classification in medical expert systems', *Theoretical Medicine* 9, 167-177.

Hume, D.: 1978, *A Treatise of Human Nature*, edited, with an analytical index, by L. A. Selby-Bigge, L.A., 2d ed., with text revised and variant readings by Nidditch, P. H., Oxford University Press, Clarendon Press, Oxford.

Hunter, K.M.: 1991, *Doctor's Stories: The Narrative Structure of Medical Knowledge*, Princeton University Press, Princeton, New Jersey.

Husserl, E.: 1970, *The Crisis of European Sciences and Transcendental Phenomenology*, translated, with an introduction by Carr, D., Northwestern University Press, Evanston, Illinois.

Illich, I.: 1976, *Medical Nemesis: The Expropriation of Health*, Pantheon Books, New York.

Jensen, U.J.: 1984, 'A critique of essentialism in medicine', in Nordenfelt, L. and Lindahl, B.I.B (eds.), *Health, Disease, and Causal Explanations in Medicine*, D. Reidel Publishing Co., Dordrecht, The Netherlands pp. 63-73.

Johnson, P.E.: 1983, 'What kind of expert should a system be?', *Journal of Medicine and Philosophy* 8, 77-97.

Jonas, H.: 1982, *The Phenomenon of Life: Toward a Philosophical Biology*, University of Chicago Press, Chicago.

Jordan, D., and Kopp, N.: 1994, 'Workshop: Definition and criteria for classification of lesions observed in Sudden Infant Death Syndrome for further research on the central nervous system: Proposal for a consensus', *Biology of the Neonate* 65, 272-280.

Kahneman, D., and Tversky, A.: 1982, 'Variants of uncertainty', *Cognition* 11, 143-157.

Kass, L.R.: 1975, 'Regarding the end of medicine and the pursuit of health', *Public Interest* 40, 11-42.

Keefe, R. and Smith P.: 1996, 'Introduction: Theories of Vagueness', in Keefe, R. and Smith, P. (eds.), *Vagueness: A Reader*, MIT Press, Cambridge, Massachusetts, pp. 1-57.

King, L.S.: 1954, 'What is disease?', *Philosophy of Science* 21, 193-203.

Kirkham, R.L.: 1992, *Theories of Truth: A Critical Introduction*, MIT Press, Cambridge, Massachusetts.

Kleinman, A.: 1988, *The Illness Narratives: Suffering, Healing, and the Human Condition*, Basic Books, New York.

Kleinmuntz, B.: 1968, 'The processing of clinical information by man and machine', in Kleinmuntz, B. (ed.), *Formal Representation of Human Judgment*, John Wiley & Sons, New York, pp. 149-186.

Knottnerus, J. A., Knipschild, P. G., and Sturmans, F.: 1989, 'Symptoms and selection bias: The influence of selection towards specialist care on the relationship between symptoms and diagnoses', *Theoretical Medicine* 10, 67-81.

Kripke, S.A.: 1980, *Naming and Necessity*, Harvard University Press, Cambridge, Massachusetts.

Kuhn, T.S.: 1970a, *The Structure of Scientific Revolutions*, 2nd ed., University of Chicago Press, Chicago.

Kuhn, T.S.: 1970b, 'Reflections on my critics', in Lakatos, I. And Musgrave, A. (eds.), *Criticism and the Growth of Knowledge*, Cambridge University Press, Cambridge, England, pp. 231-278.

Kuhn, T.S.: 1977, 'Second thoughts on paradigms', in Suppe, F. (ed.), *The Structure of Scientific Theories,* 2d ed., University of Illinois Press, Urbana, Illinois, pp. 459-482.

Kuhn, T.S.: 1983, 'Commensurability, comparability, communicability', in Asquith, P.D., and Nickles, T. (eds.), *PSA 1982: Proceedings of the 1982 Biennial Meeting of the Philosophy of Science Association*, Philosophy of Science Association, East Lansing, Michigan, pp. 669-688.

Kurtz, P.W.: 1958, 'Need reduction and normal value', *Journal of Philosophy* 55, 555-568.

Laennec, R.: 1982, 'Traité d'anatomie patholigique', in *Laennec: Catalogue des manuscrits scientifiques*, ed. Boulle, L., Grmek, M.D., Lupovici, C., and Samion-Contet, J., Masson et Fondation Singer-Polignac, Paris, MS. 2186 (III).

Lane, H.L.: 1994, *The Mask of Benevolence: Disabling the Deaf Community*, Alfred A. Knopf, New York.

Laor, N. and Agassi, J.: 1990, *Diagnosis: Philosophical and Medical Perspectives.* Kluwer Academic Publishers, Dordrecht, The Netherlands.

Lauritsen, J.: 1993, *The AIDS War: Propaganda, Profiteering and Genocide from the Medical-Industrial Complex*, Asklepios, New York.

Ledley, R.S. and Lusted, L.B.: 1959, 'Reasoning foundations of medical diagnosis', *Science* 130, 9-21.

Lemos, R.M.: 1986, 'Propositions, states of affairs, and facts', *Southern Journal of Philosophy* 24, 517-530.

Lemos, R.M.: 1995, *The Nature of Value: Axiological Investigations*, University Press of Florida, Gainesville, Florida.

Lennox, J.G.: 1995, 'Health as an objective value', *Journal of Medicine and Philosophy* 20, 499-511.

Lerman, C. and Croyle, R.: 1994, 'Psychological issues in genetic testing for breast cancer susceptibility', *Archives of Internal Medicine* 154, 609-616.

Liddle, G.W. and Shute, A.M.: 1969, 'The evolution of Cushing's Syndrome as a clinical entity', *Advances in Internal Medicine* 15, 155-175.

Limerick, S.R.: 1992, 'Sudden Infant Death Syndrome in historical perspective', *Journal of Clinical Pathology* 45 (supp.), 3-6.

Longino, H.E.: 1990, *Science as Social Knowledge: Values and Objectivity in Scientific Inquiry*, Princeton University Press, Princeton, New Jersey.

Löwy, I.: 1990a, 'Medical Critique [Krytyka Lekarska]: A journal of medicine and philosophy—1897-1907', *Journal of Medicine and Philosophy* 15, 653-673.

Löwy, I.: 1990b, *The Polish School of Philosophy of Medicine: From Tytus Chałubinski (1820-1889) to Ludwik Fleck (1896-1961)*, Kluwer Academic Publishers, Dordrecht, The Netherlands.

MacIntyre, A.: 1984, *After Virtue: A Study in Moral Theory*, 2d ed., University of Notre Dame Press, Notre Dame, Indiana.

Mackie, J.L.: 1974, *The Cement of the Universe: A Study of Causation*. Clarendon Press, Oxford.

Mackie, J.L.: 1977, *Ethics: Inventing Right and Wrong*. Penguin Books, New York.

Margolis, J.: 1976, 'The concept of disease', *Journal of Medicine and Philosophy* 1, 238-255.

Masterman, M.: 1970, 'The nature of paradigm', in Lakatos, I. And Musgrave, A. (eds.), *Criticism and the Growth of Knowledge*, Cambridge University Press, Cambridge, England, pp. 59-89.

Mazoué, J.G.: 1990, 'Diagnosis without doctors', *Journal of Medicine and Philosophy* 15, 559-579.

McCullough, L.B.: 1981, 'Thought-styles, diagnosis, and concepts of disease: Commentary on Ludwik Fleck', *Journal of Medicine and Philosophy* 6, 257-263.

McCullough, L.B. and Christianson, C.E.: 1981, 'Ethical dimensions of diagnosis: A case study and analysis', *Metamedicine* 2, 129-143.

Mellor, D.H.: 1977, 'Natural kinds', *British Journal for the Philosophy of Science* 28, 299-312.

Mill, J. S.: 1973, *Utilitarianism* (1861), in *The Utilitarians*, Anchor Books, New York, pp. 399-472.

Miller, R.A.: 1990, 'Why the standard view is standard: People, not machines, understand patients' problems', *Journal of Medicine and Philosophy* 15, 581-591.

Miller, R.A., Schaffner, K.F. and Meisel, A.: 1985, 'Ethical and legal issues related to the use of computer programs in clinical medicine', *Annals of Internal Medicine* 102, 529-536.

Mills, J.L.: 1993, 'Data torturing', *New England Journal of Medicine* 329, 1196-1199.

Mitchell, E.A., Becroft, D.M.P., Byard, R.W., Berry, P.J., Fleming, P.J., Krous, H.F., Helwig-Larsen, K., and Valdes-Dapena, M.: 1994, 'Definition of the Sudden Infant Death Syndrome: Keep current definition', *British Medical Journal* 309, 607.

Moore, G.E.: 1993, *Principia Ethica*, revised ed., Cambridge University Press, Cambridge, England.

Mordacci, R.: 1995, 'Health as an analogical concept', *Journal of Medicine and Philosophy* 20, 475-497.

Moseley, R.: 1993, 'Intuition in the art and science of medicine', in Delkeskamp-Hayes, C. and Gardell Cutter, M.A. (eds.), *Science, Technology, and the Art of Medicine*, Kluwer Academic Publishers, Dordrecht, The Netherlands, pp. 211-218.

Moskowitz, A.J., Kuipers, B.J., and Kassirer, J.P.: 1988, 'Dealing with uncertainty, risks, and tradeoffs in clinical decisions: A cognitive science approach', *Annals of Internal Medicine* 108, 435-449.

Murphy, E.A.: 1976, *The Logic of Medicine*, Johns Hopkins University Press, Baltimore.

Murphy, E.A.: 1979, 'Classification and its alternatives', in Engelhardt, H.T., Jr., Spicker, S.F. and Towers, B. (eds.), *Clinical Judgment: A Critical Appraisal*, D. Reidel Publishing Co., Dordrecht, The Netherlands, pp. 59-85.

Murphy, E.A.: 1988, 'The diagnostic process, the diagnosis, and homeostasis', *Theoretical Medicine* 9, 151-166.

Murphy, E.A.: 1992, 'Critique of diagnostic formalism', in Peset, J.L. and Gracia, D. (eds.), *The Ethics of Diagnosis*, Kluwer Academic Publishers, Dordrecht, The Netherlands, pp. 255-267.

Nagel, T.: 1986, *The View from Nowhere*, Oxford University Press, New York.

Najder, Z.: 1975, *Values and Evaluations*, Oxford University Press, Clarendon Press, London.

Nelkin, D. and Tancredi, L.: 1989, *Dangerous Diagnostics: The Social Power of Biological Information*, Basic Books, New York.

Nordenfelt, L.: 1997, *Talking About Health: A Philosophical Dialogue*, Rodopi, Amsterdam, The Netherlands.

Pellegrino, E.D.: 1979, 'The anatomy of clinical judgments: Some notes on right reason and right action', in Engelhardt, H.T., Jr., Spicker, S.F. and Towers, B. (eds.), *Clinical Judgment: A Critical Appraisal*, D. Reidel Publishing Co., Dordrecht, The Netherlands, pp. 169-194.

Pellegrino, E.D.: 1992, 'Value desiderata in the logical structuring of computer diagnosis', in Peset, J.L. and Gracia, D. (eds.), *The Ethics of Diagnosis*, Kluwer Academic Publishers, Dordrecht, The Netherlands, pp. 173-195.

Pellegrino, E.D., and Thomasma, D.C.: 1981, *A Philosophical Basis of Medical Practice*, Oxford University Press, New York.

Pera, M.: 1994, *The Discourses of Science*, edited by Botsford, C., University of Chicago Press, Chicago.

Perry, R.B.: 1923, *General Theory of Value: Its Meaning and Basic Principles Construed in Terms of Interest*, Harvard University Press, Cambridge, Massachusetts, reprint 1950.

Perry, R.B.: 1968, *Realms of Value*, Greenwood Press, New York.

Phillips, E.D.: 1973, *Aspects of Greek Medicine*, St. Martin's Press, New York.

Plato: 1961, *Timaeus*, in Hamilton, E. and Cairns, H. (eds.), *The Collected Dialogues of Plato*, Princeton University Press, Princeton, New Jersey, pp. 1151-1211.

Polanyi, M.: 1969, *Knowing and Being: Essays by Michael Polanyi*, Routledge & Kegan Paul, London.

Popper, K.R.: 1992, *The Logic of Scientific Discovery*, Routledge, London.

Putnam, H.: 1962, 'What theories are not', in Nagel, E., Suppes, P., and Tarski, A. (eds.), *Logic, Methodology and Philosophy of Science: Proceedings of the 1960 International Congress*, Stanford University Press, Stanford, California, pp. 240-251.

Putnam, H.: 1975, 'The meaning of 'meaning'', in his *Mind, Language and Reality: Philosophical Papers, Volume 2*, Cambridge University Press, Cambridge, England, pp. 215-271.

Putnam, H.: 1981, *Reason, Truth and History*, Cambridge University Press, Cambridge, England.

Putnam, H.: 1987, *The Many Faces of Realism*, Open Court, La Salle, Illinois.

Putnam, H.: 1990, *Realism with a Human Face*, Harvard University Press, Cambridge, Massachusetts.

Putnam, H.: 1991, 'The 'corroboration' of theories', in Boyd, R., Gasper, R., and Trout, J.D. (eds.), *The Philosophy of Science*, MIT Press, Cambridge, Massachusetts, pp. 121-137.

Putnam, R.A.: 1985, 'Creating facts and values', *Philosophy* 60, 187-204.

Quine, W.V.O.: 1980, *From a Logical Point of View: Logico-Philosophical Essays*, 2d ed., Harvard University Press, Cambridge, Massachusetts.

Railton, P.: 1986a, 'Facts and values', *Philosophical Topics* 14, 5-31.

Railton, P.: 1986b, 'Moral realism', *Philosophical Review* 95, 163-207.

Rambaud, C., Guillenminault, C., and Campbell, P.E.: 1994, 'Definition of the Sudden Infant Death Syndrome', *British Medical Journal* 308, 1439.

Rawls, J.: 1971, *A Theory of Justice*, Harvard University Press, Belknap Press, Cambridge, Massachusetts.

Reiser, S.J.: 1978, *Medicine and the Reign of Technology*, Cambridge University Press, Cambridge, England.

Rescher, N.: 1969, *Introduction to Value Theory*, Prentice-Hall, Englewood Cliffs, New Jersey.

Rescher, N.: 1990, 'How wide is the gap between facts and values?', *Philosophy and Phenomenological Research* 50, Suppl., 297-319.

Reznek, L.: 1987, *The Nature of Disease*, Routledge & Kegan Paul, London.

Reznek, L.: 1995, 'Dis-ease about kinds: Reply to D'Amico', *Journal of Medicine and Philosophy* 20, 571-584.

Rosenberg, C.E.: 1992, 'Framing disease: Illness, society, and history', in Rosenberg, C.E. and Golden, J. (eds.), *Framing Disease: Studies in Cultural History*, Rutgers University Press, New Brunswick, New Jersey, pp. xiii-xxvi.

Ross, W.D.: 1930, *The Right and the Good*, Oxford University Press, Oxford.

Russell, B.: 1950, *The Problems of Philosophy*, Oxford University Press, London.

Russell, B.: 1985, *The Philosophy of Logical Atomism*, edited by Pear, D., Open Court, La Salle, Illinois.

Russell, B.: 1996, 'Vagueness', in Keefe, R. and Smith, P. (eds.), *Vagueness: A Reader*, MIT Press, Cambridge, Massachusetts, pp. 61-68.

Sacks, O.: 1990, *The Man Who Mistook His Wife for a Hat and Other Clinical Tales*, Harper Perennial, New York.

Sade, R.M.: 1995, 'A theory of health and disease: The objectivist-subjectivist dichotomy', *Journal of Medicine and Philosophy* 20, 513-525.

Sadegh-Zadeh, K.: 1981, 'Foundations of clinical praxiology: Part I: The relativity of medical diagnosis', *Metamedicine* 2, 183-196.

Sadegh-Zadeh, K.: 1982, 'Foundations of clinical praxiology: Part II: Categorical and conjectural diagnoses', *Metamedicine* 3, 101-114.

Sadler, J.Z.: 1997, 'Recognizing values: A descriptive-causal method for medical/scientific discourses', *Journal of Medicine and Philosophy* 22, 541-565.

Salmon, W.C.: 1968, 'The justification of inductive rules of inference', in Lakatos, I. (ed.), *The Problem of Inductive Logic*, North-Holland Publishing Co., Amsterdam, pp. 24-43.

Sauvages, F. B.: 1768: *Nosologia methodica sistens morborum classes juxta Sydenhami mentem et botanicorum ordinem*, Fratrum de Tournes, Amsterdam.

Săvulescu, G.: 1976, 'Certainty and useful approximation in medical diagnosis', *Philosophie et Logique* 20, 327-330.

Sayre-McCord, G.: 1988, 'Introduction: The many moral realisms', in Sayre-McCord, G. (ed.), *Essays on Moral Realism*, Cornell University Press, Ithaca, New York, pp. 1-23.

Scadding, J.G.: 1967, 'Diagnosis: The clinician and the computer', *Lancet* 2, 877-882.

Schaffner, K.F.: 1981, 'Modeling medical diagnosis: Logical and computer approaches', *Synthese* 47, 163-199.

Schaffner, K.F.: 1985, 'Introduction', in Schaffner, K.F. (ed.), *Logic of Discovery and Diagnosis in Medicine*, University of California Press, Berkeley, pp. 1-32.

Schaffner, K.F.: 1992, 'Problems in computer diagnosis', in Peset, J.L. and Gracia, D. (eds.), *The Ethics of Diagnosis*, Kluwer Academic Publishers, Dordrecht, The Netherlands, pp. 197-241.

Schaffner, K.F.: 1993, *Discovery and Explanation in Biology and Medicine*, University of Chicago Press, Chicago.

Scheler, M.: 1973, *Formalism in Ethics and Non-Formal Ethics of Values; A New Attempt toward the Foundation of an Ethical Personalism*, translated by Frings, M. S. and. Funk, R. L., Northwestern University Press, Evanston, Illinois.

Schwartz, S., and Griffin, T.: 1986, *Medical Thinking: The Psychology of Medical Judgment and Decision Making*, Springer-Verlag, New York.

Searle, J.R.: 1991, 'Is the brain's mind a computer program?', in Hoy, R.C. and Oaklander, N.C. (eds.), *Metaphysics: Classical and Contemporary Readings*, Wadsworth Publishing Co., Belmont, California, pp. 276-285.

Sellars, W.: 1991, *Science, Perception and Reality*, Ridgeview Publishing Co., Atascadero, California.

Shorter, J.M.: 1962, 'Facts, logical atomism and reducibility', *Australasian Journal of Philosophy* 40, 283-302.

Sisson, J.C., Donnelly, M.B., Hess, G.E., and Woolliscroft, J.O.: 1991, 'The characteristics of early diagnostic hypotheses generated by physicians (experts) and students (novices) at one medical school', *Academic Medicine* 66, 607-612.

Sober, E.: 1979, 'The art and science of clinical judgment: An informational approach', in Engelhardt, H.T., Jr., Spicker, S.F. and Towers, B. (eds.), *Clinical Judgment: A Critical Appraisal*, D. Reidel Publishing Co., Dordrecht, The Netherlands, pp. 29-44.

Sonntag, S.: 1979, *Illness as Metaphor*, Vintage Books, New York.

Sosa, E. and Tooley M., eds.: 1993, *Causation*, Oxford University Press, Oxford.

Spicker, S.F.: 1993, 'Intuition and the process of medical diagnosis: The quest for explicit knowledge in the technological era', in Delkeskamp-Hayes, C. and Gardell Cutter, M.A. (eds.), *Science, Technology, and the Art of Medicine*, Kluwer Academic Publishers, Dordrecht, The Netherlands, pp. 199-210.

Stack, G.J.: 1969, 'Values and facts', *Journal of Value Inquiry* 3, 205-216.

Stack, G.J.: 1973, 'Subjectivity, facts, and values', *Studies in Philosophy and the History of Philosophy* 6, 77-120.

Statland, B.E., Winkel, P., Burke, M.D., and Galen, R.S.: 1979, 'Quantitative approaches used in evaluating laboratory measurements and other clinical data', in Henry, J.B. (ed.), *Clinical Diagnosis and Management by Laboratory Methods*, 16th ed., W. B. Saunders, Philadelphia, pp. 525-555.

Stedman, T.L.: 1995, *Stedman's Medical Dictionary*, 26th ed., Williams & Wilkins, Baltimore.

Strawson, P.F.: 1950, 'Truth', *Proceedings of the Aristotelian Society* 24 supp., 129-156.

Suppe, F.: 1973, 'Facts and empirical truth', *Canadian Journal of Philosophy* 3, 197-212.

Suppe, F.: 1977, 'The search for philosophic understanding of scientific theories', in Suppe, F. (ed.), *The Structure of Scientific Theories,* 2d ed., University of Illinois Press, Urbana, Illinois, pp. 3-241.

Swinburne, R.G.: 1961, 'Three types of thesis about fact and value', *Philosophical Quarterly* 11, 301-307.

Sydenham, T.: 1979, *The Works of Thomas Sydenham, M. D.*, from the Latin edition of Dr. Greenhill, with a life of the author by Latham, R.G., facsimile of the Sydenham Society's two-volume edition of 1848-1850, Classics of Medicine Library, Birmingham, Alabama.

Szasz, T.S.: 1970, *Ideology and Insanity: Essays on the Psychiatric Dehumanization of Man*, Anchor Books, Garden City, New York.

Szolovits, P. and Pauker, S.G.: 1978, 'Categorical and probabilistic reasoning in medical diagnosis', *Artificial Intelligence* 11, 115-144.

Taylor, F.K.: 1979, *The Concepts of Illness, Disease and Morbus*, Cambridge University Press, Cambridge, England.

Taylor, F.K.: 1981, 'Disease concepts and the logic of classes', *British Journal of Medical Psychology* 54, 277-286.

Temkin, O.: 1961, 'The scientific approach to disease: Specific entity and individual sickness', in Crombie, A.C. (ed.), *Scientific Change: Historical Studies in the Intellectual, Social and Technical Conditions for Scientific Discovery and Technical Invention, From Antiquity to the Present: Symposium on the History of Science, University of Oxford, 9-15 July 1961*, Basic Books, New York, pp. 628-647.

Temkin, O.: 1973, 'Health and disease', in Wiener, P. (ed.), *Dictionary of the History of Ideas: Studies of Selected Pivotal Ideas*, Scribner's, New York.

Tooms, S.K.: 1992, *The Meaning of Illness: A Phenomenological Account of the Different Perspectives of Physician and Patient*, Kluwer Academic Publishers, Dordrecht, The Netherlands.

Toon, P.D.: 1981, 'Defining 'disease': Classification must be distinguished from evaluation', *Journal of Medical Ethics* 7, 197-201.

Toulmin, S.: 1970, 'The nature of paradigm', in Lakatos, I. And Musgrave, A. (eds.), *Criticism and the Growth of Knowledge*, Cambridge University Press, Cambridge, England, pp. 39-47.

Trenn, T.J.: 1981, 'Ludwik Fleck's 'On the foundations of medical knowledge'', *Journal of Medicine and Philosophy* 6, 237-256.

Tumulty, P.: 1988, 'A contemporary bridge from facts to values: But will natural law theorists pay the toll?', *International Philosophical Quarterly* 28, 53-63.

Vácha, J.: 1978, 'Biology and the problem of normality', *Scientia* 113, 823-865.

Vácha, J.: 1985, 'German constitutional doctrine in the 1920s and 1930s and pitfalls of the contemporary conception of normality in biology and medicine', *Journal of Medicine and Philosophy* 10, 339-367.

Valdes-Dapena, M.: 1992, 'The Sudden Infant Death Syndrome: Pathologic findings', *Clinics in Perinatology* 19, 701-716.

Veatch, R.M.: 1973a, 'Generalization of expertise: Scientific expertise and value judgments', *Hastings Center Studies* 1, 29-40.

Veatch, R.M.: 1973b, 'The medical model: Its nature and problems', *Hastings Center Studies* 1, 59-76.

Veatch, R.M.: 1976, *Value-Freedom in Science and Technology: A Study of the Importance of the Religious, Ethical, and Other Socio-Cultural Factors in Selected Medical Decisions Regarding Birth Control*, Scholars Press, Missoula, Montana.

Veatch, R.M. and Stempsey, W.E.: 1995, 'Incommensurability: Its implications for the patient/physician relation', *Journal of Medicine and Philosophy* 20, 253-269.

Vineis, P.: 1993, 'Definition and classification of cancer: Monothetic or polythetic?', *Theoretical Medicine* 14, 249-256.

Virchow, R.L.K.: 1958, *Disease, Life, and Man: Selected Essays*, Stanford University Press, Stanford, California.

Vision, G.: 1988, *Modern Anti-Realism and Manufactured Truth*, Routledge, London.

Walhout, D.: 1978, *The Good and the Realm of Values*, University of Notre Dame Press, Notre Dame, Indiana.

Wartofsky, M.W.: 1986, 'Clinical judgment, expert programs, and cognitive style: A counter-essay in the logic of diagnosis', *Journal of Medicine and Philosophy* 11, 81-92.

Weber, E.U., Bockenholt, U., Hilton, D.J., and Wallace, B.: 1993, 'Determinants of diagnostic hypothesis generation: Effects of information, base rates, and experience', *Journal of Experimental Psychology: Learning, Memory, and Cognition* 19, 1151-1164.

Whitbeck, C.: 1978, 'Four basic concepts of medical science', in Asquish, P.D. and Hacking, I. (eds.), *PSA 1978: Proceedings of the 1978 Biennial Meeting of the Philosophy of Science Association*, Philosophy of Science Association, East Lansing, Michigan, pp. 210-222.

Whitbeck, C.: 1981a, 'A theory of health', in Caplan, A.L., Engelhardt, H.T., Jr., and McCartney, J.J (eds.), *Concepts of Health and Disease: Interdisciplinary Perspectives*, Addison-Wesley Publishing Co., Reading, Massachusetts, pp. 611-626.

Whitbeck, C.: 1981b, 'What is diagnosis? Some critical reflections', *Metamedicine* 2, 319-329.

Willard, L.D.: 1982, 'Needs and medicine', *Journal of Medicine and Philosophy* 7, 259-274.

Willinger, M., James, L.S., and Catz, C.: 1991, 'Defining the Sudden Infant Death Syndrome (SIDS): Deliberations of an expert panel convened by the National Institute of Child Health and Human Development', *Pediatric Pathology* 11, 677-684.

Wittgenstein, L.: 1973, *Philosophical Investigations: The English Text of the Third Edition*, translated by Anscome, G.E.M., Macmillan Publishing Co., New York.

World Health Organization: 1977, *International Classification of Diseases. Manual of the International Statistical Classification of Diseases, Injuries, and Causes of Death: Based on the Recommendations of the Ninth Revision Conference, 1975, and Adopted by the Twenty-ninth World Health Assembly*, World Health Organization, Geneva, Switzerland.

Wright, C.: 1996, 'Language-mastery and the sorites paradox', in Keefe, R. and Smith, P. (eds.), *Vagueness: A Reader*, MIT Press, Cambridge, Massachusetts, pp. 151-173.

Wright, L.: 1973, 'Functions', *Philosophical Review* 82, 139-168.

Wulff, H.R.: 1976, *Rational Diagnosis and Treatment*, Blackwell Scientific Publications, Oxford.

Wulff, H.R.: 1992, 'Computers and clinical thinking', in Peset, J.L. and Gracia, D. (eds.), *The Ethics of Diagnosis*, Kluwer Academic Publishers, Dordrecht, The Netherlands, pp. 243-254.

Ziporyn, T.D.:, 1992, *Nameless Diseases*, Rutgers University Press, New Brunswick, New Jersey.

INDEX

Philosophy and Medicine

1. H. Tristram Engelhardt, Jr. and S.F. Spicker (eds.): *Evaluation and Explanation in the Biomedical Sciences.* 1975   ISBN 90-277-0553-4
2. S.F. Spicker and H. Tristram Engelhardt, Jr. (eds.): *Philosophical Dimensions of the Neuro-Medical Sciences.* 1976   ISBN 90-277-0672-7
3. S.F. Spicker and H. Tristram Engelhardt, Jr. (eds.): *Philosophical Medical Ethics.* Its Nature and Significance. 1977   ISBN 90-277-0772-3
4. H. Tristram Engelhardt, Jr. and S.F. Spicker (eds.): *Mental Health.* Philosophical Perspectives. 1978   ISBN 90-277-0828-2
5. B.A. Brody and H. Tristram Engelhardt, Jr. (eds.): *Mental Illness.* Law and Public Policy. 1980   ISBN 90-277-1057-0
6. H. Tristram Engelhardt, Jr., S.F. Spicker and B. Towers (eds.): *Clinical Judgment.* A Critical Appraisal. 1979   ISBN 90-277-0952-1
7. S.F. Spicker (ed.): *Organism, Medicine, and Metaphysics.* Essays in Honor of Hans Jonas on His 75th Birthday. 1978   ISBN 90-277-0823-1
8. E.E. Shelp (ed.): *Justice and Health Care.* 1981
                    ISBN 90-277-1207-7; Pb 90-277-1251-4
9. S.F. Spicker, J.M. Healey, Jr. and H. Tristram Engelhardt, Jr. (eds.): *The Law-Medicine Relation.* A Philosophical Exploration. 1981   ISBN 90-277-1217-4
10. W.B. Bondeson, H. Tristram Engelhardt, Jr., S.F. Spicker and J.M. White, Jr. (eds.): *New Knowledge in the Biomedical Sciences.* Some Moral Implications of Its Acquisition, Possession, and Use. 1982   ISBN 90-277-1319-7
11. E.E. Shelp (ed.): *Beneficence and Health Care.* 1982   ISBN 90-277-1377-4
12. G.J. Agich (ed.): *Responsibility in Health Care.* 1982   ISBN 90-277-1417-7
13. W.B. Bondeson, H. Tristram Engelhardt, Jr., S.F. Spicker and D.H. Winship: *Abortion and the Status of the Fetus.* 2nd printing, 1984   ISBN 90-277-1493-2
14. E.E. Shelp (ed.): *The Clinical Encounter.* The Moral Fabric of the Patient-Physician Relationship. 1983   ISBN 90-277-1593-9
15. L. Kopelman and J.C. Moskop (eds.): *Ethics and Mental Retardation.* 1984
                    ISBN 90-277-1630-7
16. L. Nordenfelt and B.I.B. Lindahl (eds.): *Health, Disease, and Causal Explanations in Medicine.* 1984   ISBN 90-277-1660-9
17. E.E. Shelp (ed.): *Virtue and Medicine.* Explorations in the Character of Medicine. 1985   ISBN 90-277-1808-3
18. P. Carrick: *Medical Ethics in Antiquity.* Philosophical Perspectives on Abortion and Euthanasia. 1985   ISBN 90-277-1825-3; Pb 90-277-1915-2
19. J.C. Moskop and L. Kopelman (eds.): *Ethics and Critical Care Medicine.* 1985
                    ISBN 90-277-1820-2
20. E.E. Shelp (ed.): *Theology and Bioethics.* Exploring the Foundations and Frontiers. 1985   ISBN 90-277-1857-1

Philosophy and Medicine

21. G.J. Agich and C.E. Begley (eds.): *The Price of Health.* 1986
ISBN 90-277-2285-4
22. E.E. Shelp (ed.): *Sexuality and Medicine.* Vol. I: Conceptual Roots. 1987
ISBN 90-277-2290-0; Pb 90-277-2386-9
23. E.E. Shelp (ed.): *Sexuality and Medicine.* Vol. II: Ethical Viewpoints in Transition.
1987 ISBN 1-55608-013-1; Pb 1-55608-016-6
24. R.C. McMillan, H. Tristram Engelhardt, Jr., and S.F. Spicker (eds.): *Euthanasia
and the Newborn.* Conflicts Regarding Saving Lives. 1987
ISBN 90-277-2299-4; Pb 1-55608-039-5
25. S.F. Spicker, S.R. Ingman and I.R. Lawson (eds.): *Ethical Dimensions of Geriatric
Care.* Value Conflicts for the 21th Century. 1987 ISBN 1-55608-027-1
26. L. Nordenfelt: *On the Nature of Health.* An Action-Theoretic Approach. 2nd,
rev. ed. 1995 SBN 0-7923-3369-1; Pb 0-7923-3470-1
27. S.F. Spicker, W.B. Bondeson and H. Tristram Engelhardt, Jr. (eds.): *The Contra-
ceptive Ethos.* Reproductive Rights and Responsibilities. 1987
ISBN 1-55608-035-2
28. S.F. Spicker, I. Alon, A. de Vries and H. Tristram Engelhardt, Jr. (eds.): *The Use
of Human Beings in Research.* With Special Reference to Clinical Trials. 1988
ISBN 1-55608-043-3
29. N.M.P. King, L.R. Churchill and A.W. Cross (eds.): *The Physician as Captain of
the Ship.* A Critical Reappraisal. 1988 ISBN 1-55608-044-1
30. H.-M. Sass and R.U. Massey (eds.): *Health Care Systems.* Moral Conflicts in
European and American Public Policy. 1988 ISBN 1-55608-045-X
31. R.M. Zaner (ed.): *Death: Beyond Whole-Brain Criteria.* 1988
ISBN 1-55608-053-0
32. B.A. Brody (ed.): *Moral Theory and Moral Judgments in Medical Ethics.* 1988
ISBN 1-55608-060-3
33. L.M. Kopelman and J.C. Moskop (eds.): *Children and Health Care.* Moral and
Social Issues. 1989 ISBN 1-55608-078-6
34. E.D. Pellegrino, J.P. Langan and J. Collins Harvey (eds.): *Catholic Perspectives
on Medical Morals.* Foundational Issues. 1989 ISBN 1-55608-083-2
35. B.A. Brody (ed.): *Suicide and Euthanasia.* Historical and Contemporary Themes.
1989 ISBN 0-7923-0106-4
36. H.A.M.J. ten Have, G.K. Kimsma and S.F. Spicker (eds.): *The Growth of Medical
Knowledge.* 1990 ISBN 0-7923-0736-4
37. I. Löwy (ed.): *The Polish School of Philosophy of Medicine.* From Tytus
Chałubiński (1820–1889) to Ludwik Fleck (1896–1961). 1990
ISBN 0-7923-0958-8
38. T.J. Bole III and W.B. Bondeson: *Rights to Health Care.* 1991
ISBN 0-7923-1137-X

Philosophy and Medicine

39. M.A.G. Cutter and E.E. Shelp (eds.): *Competency. A Study of Informal Competency Determinations in Primary Care*. 1991   ISBN 0-7923-1304-6
40. J.L. Peset and D. Gracia (eds.): *The Ethics of Diagnosis*. 1992
     ISBN 0-7923-1544-8
41. K.W. Wildes, S.J., F. Abel, S.J. and J.C. Harvey (eds.): *Birth, Suffering, and Death*. Catholic Perspectives at the Edges of Life. 1992 [CSiB-1]
     ISBN 0-7923-1547-2; Pb 0-7923-2545-1
42. S.K. Toombs: *The Meaning of Illness*. A Phenomenological Account of the Different Perspectives of Physician and Patient. 1992
     ISBN 0-7923-1570-7; Pb 0-7923-2443-9
43. D. Leder (ed.): *The Body in Medical Thought and Practice*. 1992
     ISBN 0-7923-1657-6
44. C. Delkeskamp-Hayes and M.A.G. Cutter (eds.): *Science, Technology, and the Art of Medicine*. European-American Dialogues. 1993   ISBN 0-7923-1869-2
45. R. Baker, D. Porter and R. Porter (eds.): *The Codification of Medical Morality*. Historical and Philosophical Studies of the Formalization of Western Medical Morality in the 18th and 19th Centuries, Volume One: Medical Ethics and Etiquette in the 18th Century. 1993   ISBN 0-7923-1921-4
46. K. Bayertz (ed.): *The Concept of Moral Consensus*. The Case of Technological Interventions in Human Reproduction. 1994   ISBN 0-7923-2615-6
47. L. Nordenfelt (ed.): *Concepts and Measurement of Quality of Life in Health Care*. 1994 [ESiP-1]   ISBN 0-7923-2824-8
48. R. Baker and M.A. Strosberg (eds.) with the assistance of J. Bynum: *Legislating Medical Ethics*. A Study of the New York State Do-Not-Resuscitate Law. 1995
     ISBN 0-7923-2995-3
49. R. Baker (ed.): *The Codification of Medical Morality*. Historical and Philosophical Studies of the Formalization of Western Morality in the 18th and 19th Centuries, Volume Two: Anglo-American Medical Ethics and Medical Jurisprudence in the 19th Century. 1995   ISBN 0-7923-3528-7; Pb 0-7923-3529-5
50. R.A. Carson and C.R. Burns (eds.): *Philosophy of Medicine and Bioethics*. A Twenty-Year Retrospective and Critical Appraisal. 1997 ISBN 0-7923-3545-7
51. K.W. Wildes, S.J. (ed.): *Critical Choices and Critical Care*. Catholic Perspectives on Allocating Resources in Intensive Care Medicine. 1995 [CSiB-2]
     ISBN 0-7923-3382-9
52. K. Bayertz (ed.): *Sanctity of Life and Human Dignity*. 1996
     ISBN 0-7923-3739-5
53. Kevin Wm. Wildes, S.J. (ed.): *Infertility: A Crossroad of Faith, Medicine, and Technology*. 1996   ISBN 0-7923-4061-2
54. Kazumasa Hoshino (ed.): *Japanese and Western Bioethics*. Studies in Moral Diversity. 1996   ISBN 0-7923-4112-0

Philosophy and Medicine

55. E. Agius and S. Busuttil (eds.): *Germ-Line Intervention and our Responsibilities to Future Generations*. 1998   ISBN 0-7923-4828-1

56. L.B. McCullough: *John Gregory and the Invention of Professional Medical Ethics and the Professional Medical Ethics and the Profession of Medicine*. 1998
     ISBN 0-7923-4917-2

57. L.B. McCullough: *John Gregory's Writing on Medical Ethics and Philosophy of Medicine*. 1998 [CiME-1]   ISBN 0-7923-5000-6

58. H.A.M.J. ten Have and H.-M. Sass (eds.): *Consensus Formation in Healthcare Ethics*. 1998 [ESiP-2]   ISBN 0-7923-4944-X

59. H.A.M.J. ten Have and J.V.M. Welie (eds.): *Ownership of the Human Body*. Philosophical Considerations on the Use of the Human Body and its Parts in Healthcare. 1998 [ESiP-3]   ISBN 0-7923-5150-9

60. M.J. Cherry (ed.): *Persons and Their Bodies*. Rights, Responsibilities, Relationships. 1999   ISBN 0-7923-5701-9

61. R. Fan (ed.): *Confucian Bioethics*. 1999 [APSiB-1]   ISBN 0-7923-5853-8

62. L.M. Kopelman (ed.): *Building Bioethics*. Conversations with Clouser and Friends on Medical Ethics. 1999   ISBN 0-7923-5853-8

63. W.E. Stempsey: *Disease and Diagnosis*. 2000   PB ISBN 0-7923-6322-1

64. H.T. Engelhardt (ed.): *The Philosophy of Medicine*. Framing the Field. 2000
     ISBN 0-7923-6223-3

65. S. Wear, J.J. Bono, G. Logue and A. McEvoy (eds.): *Ethical Issues in Health Care on the Frontiers of the Twenty-First Century*. 2000   ISBN 0-7923-6277-2

KLUWER ACADEMIC PUBLISHERS – DORDRECHT / BOSTON / LONDON